THE CHALLENGE OF ENVIRONMENTAL MANAGEMENT IN URBAN AREAS

The Challenge of Environmental Management in Urban Areas

Edited by
ADRIAN ATKINSON
JULIO D. DÁVILA
EDÉSIO FERNANDES
MICHAEL MATTINGLY

Ashgate

Aldershot • Brookfield USA • Singapore • Sydney

Published by
Ashgate Publishing Ltd
Gower House
Croft Road
Aldershot
Hants GU11 3HR
England

Ashgate Publishing Company
Old Post Road
Brookfield
Vermont 05036
USA

British Library Cataloguing in Publication Data
The challenge of environmental management in urban areas. -
    (Ashgate studies in environmental policy and practice)
    1. Urban ecology 2. Environmental management 3. Land use,
    Urban - Cross-cultural studies 4. Urban policy -
    Environmental aspects
    I. Atkinson, Adrian, 1943-
    333.7'7'16

Library of Congress Catalog Card Number: 99-72605

ISBN 1 84014 525 0

Printed and bound by Athenaeum Press, Ltd.,
Gateshead, Tyne & Wear.

Contents

PART III: ORGANISATION AND POLITICS

Acknowledgements

This book contains an edited selection of papers presented at the international conference entitled 'The Challenge of Environmental Management in Metropolitan Areas' which took place at the University of London on 19-20 June 1997. The conference was jointly organised by the Development Planning Unit (DPU), University College London, and the Institute of Commonwealth Studies, University of London. The editors would like to thank all the participants in the conference, particularly those who presented papers and shared their valuable experiences with the other speakers and participants, as well as the workshop convenors and rapporteurs who provided the plenary sessions with very useful feedback on their group's deliberations.

At the Institute of Commonwealth Studies, we would like to thank Professor James Manor, Ms Sonja Jansen, Ms Imelda McGowan and Mrs Rowena Rochanowska, for their support and help. At the DPU, we would like to thank Ms Sue Raychadhuri, Mr Frankie Liew, Mr Stephane Lambert and Mr Enrico Corubolo.

We are also indebted to the Citicorp Foundation and Citibank London for their generous grants, which made the conference possible. We would especially like to thank Mr Peter Thorpe, Mr Robert Annibale, and Ms Kathryn Carassalini.

Contributors

Claudio Acioly Jr. Institute for Housing and Urban Development Studies, IHS, P.O. Box 1935, 3000 BX, Rotterdam, The Netherlands. E-mail: clac@admin.ihs.nl

Adriana Allen. Development Planning Unit, University College London, 9 Endsleigh Gardens, London WC1H 0ED, United Kingdom. E-mail: a.allen@ucl.ac.uk

P. B. Anand. Development and Project Planning Centre, University of Bradford, Bradford BD7 1DP, United Kingdom.

Adrian Atkinson. Development Planning Unit, University College London, 9 Endsleigh Gardens, London WC1H 0ED, United Kingdom. E-mail: adrian.atkinson@ucl.ac.uk

Sergio F. León Balza. Instituto de Estudios Urbanos, Universidad Católica de Chile, El Comendador 1916, Santiago, Chile. Tel. +56 (2) 686 5504; fax: +56 (2) 232 8805.

Kalyan Biswas. Calcutta Environmental Management Strategy and Action Plan, Dept. of Environment, Government of West Bengal, 17A Everest House, 46C Chowringhee Road, Calcutta 700 071, India. Tel. +91 (33) 282 4504; fax +91 (33) 282 7021.

Chris Church. PO Box 893, London E5 9RU, United Kingdom. E-mail: cjchurch@geo2.poptel.org.uk

Jeff Cooper. Producer Responsibility Registration Unit, Environment Agency, Wah Kwong House, 10/11 Albert Embankment, London SE1 7SP, United Kingdom. Tel: +44 (181) 305 4036; fax: +44 (181) 305 4027 (please note that the '181' code will change to '208' from 22 April 2000).

Julio D. Dávila. Development Planning Unit, University College London, 9 Endsleigh Gardens, London WC1H 0ED, United Kingdom. E-mail: j.davila@ucl.ac.uk

Ben K. Doe. Accra Sustainable Programme, P. O. Box 2892, Accra, Ghana. E-mail: asp@ncs.com.gh

David J. Edelman. School of Planning, College of Design, Architecture, Art & Planning, Aronoff Center for Design and Art, University of Cincinnati, P.O. Box 210016, Cincinnati, OH 45221-0016, USA. E-mail: d.edelman@uc.edu

Edésio Fernandes. Institute of Commonwealth Studies, 22 Russell Square, London WC1B 5DS. E-mail: edesiofernandes@compuserve.com

Osvaldo Girardin. Fundación Bariloche, Programa de Medio Ambiente y Desarrollo (MEDA/FB), Piedras 482-2H, (1070) Buenos Aires, Argentina. E-mail: ideefb@mbox.servicenet.com.ar

Nelson Gouveia. Centro de Estudos de Cultura Contemporanea-CEDEC, Rua Airosa Galvao, 64, Sao Paulo - 05002-070 - SP, Brazil. E-mail: ngouveia@mandic.com.br

Gabriela Greco. University of Buenos Aires, Campichuelo 147 8C, (1405) Buenos Aires, Argentina. E-mail: ideefb@mbox.servicenet.com.ar

Simon Guy. Centre for Urban Technology, Department of Town and Country Planning, University of Newcastle, Newcastle NE1 7RU, United Kingdom. E-mail: s.c.guy@ncl.ac.uk

John Handmer. School of Social Science, Middlesex University, Queensway, Enfield, Middlesex, EN3 4SF, United Kingdom. E-mail: j.handmer@mdx.ac.uk

Pedro Jacobi. Programa de Pós-Graduação em Ciência Ambiental, Universidade de São Paulo, Rua do Anfiteatro 181 -Colméias/Favo 15, 05508-900 Cidade Universitária, São Paulo, Brazil. E-mail: pjacobi@sysnetway.com.br

J. M. Lusugga Kironde. University College of Lands and Architectural Studies, PO Box 35176, Dar es Salaam, Tanzania. E-mail: ihsbr@ud.co.tz

Alphonce G. Kyessi. Institute of Housing Studies and Building Research, PO Box 35124, Dar es Salaam, Tanzania. E-mail: kyessi@unidar.gn.apc.org

Simon Marvin. Centre for Urban Technology, Department of Town and Country Planning, University of Newcastle, Newcastle NE1 7RU, United Kingdom. E-mail: s.j.marvin@ncl.ac.uk

Michael Mattingly. Development Planning Unit, University College London, 9 Endsleigh Gardens, London WC1H 0ED, United Kingdom. E-mail: m.mattingly@ucl.ac.uk

Margarita Pacheco. Associate professor, Instituto de Estudios Ambientales, Universidad Nacional de Colombia, AA 14490, Santafé de Bogotá, Colombia. Contact address: 54 rue Liotard, 1202 Genève, Switzerland. E-mail: mpacheco@iprolink.fr

Paul Procee. Department of Urban Environmental Management, Institute for Housing and Urban Development Studies (IHS), P.O. Box 1935, 3000 BX Rotterdam, The Netherlands. E-mail: papr@admin.ihs.nl.

V. U. Ratnayake. Wimala Estate, Nugegoda, Sri Lanka. E-mail: iprp@comtech.slt.lk

Celina Souza. Department of Finance and Public Policies, Federal University of Bahia, Rua Quintino de Carvalho 153/702, Salvador, Bahia, Brazil. E-mail celina@ufba.br

Carolyn Stephens. London School of Hygiene & Tropical Medicine, Keppel Street, London WC1E 7HT, United Kingdom. E-mail: c.stephens@lshtm.ac.uk.

Luz Stella Velásquez. Associate professor, Instituto de Estudios Ambientales, Universidad Nacional de Colombia, El Cable, Manizales, Colombia. E-mail: bioluz@emtelsa.com.co

Doris Tetteh. Accra Sustainable Programme, P. O. Box 2892, Accra, Ghana. E-mail: asp@ncs.com.gh

Ayse Yonder. Graduate Center for Planning and the Environment, Pratt Institute, 200 Willoughby Avenue, Brooklyn, New York 11205, USA. E-mail: ayonder@ix.netcom.com

Amanda Younge. Director of Planning. Civic Centre, 12 Hertzog Boulevard, P.O. Box 1694, Cape Town 8000, Republic of South Africa. E-mail: ayounge@ctcc.gov.za

xiii

1 The Challenge of Environmental Management in Urban Areas: An Introduction

ADRIAN ATKINSON AND JULIO D. DÁVILA

Environmental Impacts of Urbanisation

'The environment' arose as an issue of major political and academic interest already in the early 1970s. Since the mid 1980s, this interest has taken on a greater importance and broadened out to a concern for 'sustainable development'.[1] However, the contribution of cities to the creation of environmental problems and the role which they might play in mitigating these problems - or even in being a significant context for solving the problems - has only arisen since the late 1980s. Nevertheless, over the past few years a significant literature both on the urban environment, conventionally seen, and on the role of cities in achieving sustainable development has arisen. This book goes beyond the debates about approach and policy options to look at what is taking place. Over twenty cases are presented which give a unique breadth to existing knowledge of environmental management practice in urban areas.

In practice, 'environmental concern' means very different things to different people. Concern for environmental problems in the rapidly growing cities of the South has occupied a particular set of interests: this naturally includes those who live in, or have a responsibility for managing, the cities themselves and then it includes academics and others involved in issues of urbanisation in the countries of the South.

Within this milieu there has tended to be a presumption that urban environmental problems are unique to the cities of the South and that cities in the North present a model of the solution to these problems. However, those involved in urban issues in Northern cities know otherwise: there is a strong and growing movement - particularly revolving around the generation of Local Agenda 21 programmes discussed further below - that asserts that the

1

management of Northern cities is also in urgent need of revision to take adequate account of environmental problems. The focus of concern is, however, somewhat different.

In the case of Southern cities the focus of attention has hitherto been almost entirely upon the inability of urban authorities to implement - often well understood and quite straightforward - measures to control the negative environmental side-effects of rapid urbanisation. Known as the environmental 'brown agenda', this involves a concern to devise appropriate methods to mitigate the health and efficiency impacts of air and water pollution and to improve the supply of basic infrastructural services, including water, solid waste disposal and transport, and encompassing also social services. The poor sanitary conditions and relatively high price paid by the poor for urban services due to inefficiencies in their provision in the cities of the South are well known and well documented (Hardoy et al, 1992).

In the case of the Northern countries, the new urban environmental concern is more related to a growing awareness of the unsustainability of the development process as a whole, including the way in which cities are built and function; this is referred to as the 'green agenda'. In fact most of the new literature on urbanisation in the South focuses attention on the processes leading to the rapid growth of cities and on the results that these are generation. With respect to the urban environment in these cities, the urgency of brown agenda problems become the major focus of attention. The question of sustainability appears to be a luxury that Southerners cannot afford until the people have overcome poverty and the cities are as well-managed as those on the North.

However, the sustainability problem is not merely a question of Northern cities. Southern cities too are in many respects unsustainable even while they contain so much poverty and suffer from so many brown agenda problems. It is clearly necessary to ensure that a focus on environmental problems of Southern cities should take account not only of immediate problems and how these might be alleviated but that the solutions to these are sustainable into the future (Haughton and Hunter, 1994; White, 1994; Satterthwaite, 1997).

Although the term 'sustainable development' is now common currency, its meaning and possible application to urban management remain clouded in confusions and contradictory interpretations. The primary meaning of the term is one which refers to the fact that globally - and often more obviously locally - resources are being depleted with the possibility that future generations will not be able to sustain themselves. Often this is a question of degrading resources through pollution and through the inappropriate disposal of wastes. But more usually it is a matter of mis-management and mis-organisation of the use of resources. This has been referred to as 'secondary sustainability' (cfr. paper by Allen in this book).[2]

Thus sustainability is not only about the continuing supply of resources but also about social and economic systems, about politics and culture. At this level, the brown and green agendas have a common concern and focus. The danger of discussions in this area are, however, that one must, *prima facie*, question whether sustaining social and political - or even cultural systems - is a good idea or whether environmental sustainability is precisely caused by the continuation of these practices (Hardoy et al, 1992).

The primary focus of the debate around secondary sustainability is therefore about the capacity for change: concerning raising awareness of the need for change in personal and social practices and the capacity of those in responsible positions (such as urban authorities) to propose and then carry out the necessary actions to ensure urban sustainability. Thus the brown agenda cannot be separated from the green agenda: improvement in urban management must involve improving planning and management of the future sustainability of urban regions at the same time as improving the environmental conditions in which much of the urban population is living and working in the cities of the South.

Administering the Management of the Metropolitan Environment

The growing interest in urban environment in recent years has resulted in initiatives in specific areas whilst others have been given less attention. For instance there has been a major focus on local community involvement in environmental management. These have given us already substantial experience in participatory projects which are in most cases, however, concerned with very local problems in specific communities (Gilbert et al, 1996; Pugh, 1996).

On the other hand, very little attention has been focused on environmental management of towns and cities as a whole and on their metropolitan regions that might include the impacts of urbanisation on rural areas well beyond the limits of the built city. In considering the problematic of sustainability of cities it would seem that the resource interactions between cities and their regions and the administration of environmental resources on the part of city-wide authorities are crucial issues. The papers collected in this book were initially produced for a conference aiming to survey environmental management at a metropolitan level. In the end, that conference produced no evidence that metropolitan level environmental management was actually taking place. This was a significant finding from more than 30 accounts of practice in cities all over the world.

The question of administering metropolitan areas in order to ensure their accommodating the needs of the inhabitants adequately - and now also with an eye to sustainability - is not a simple question. The modes of administration are many and varied and it is clear from the deteriorating economic and environmental circumstances of large numbers of urban inhabitants that many of the current solutions are inadequate to the task.

The main issues arising are as follows. An administrative arrangement adequate to a small town is inevitably inadequate for a city. Unfortunately a smooth scaling up of management systems as towns grow into cities is not just a matter of hiring more personnel. Urban boundaries cannot simply be changed each year. Structures of urban governance are difficult to change and particularly difficult to establish in a way that will anticipate future developments - especially in a rapidly changing world where things anticipated rarely materialise as expected.

But beyond the technical, there are issues of the conception of who should be doing what. In most countries of the South, the presumption has been that the central government was the only competent authority and that it should take the main responsibility for all aspects of governance, including the local. There was some move in the face of burgeoning problems of urbanisation in the 1960s to plan for regional development and to attempt to establish metropolitan authorities in many urbanising regions of the South. These were, however, largely unsuccessful for a number of reasons.[3] Thus urban management remained in the hands of often fragmented local authorities (cities and urban regions made up of many small municipalities that cooperated very poorly with one another).

The major reason for the failure of metropolitan governance has been political: the unwillingness of existing authorities to relinquish powers. This is not a new phenomenon, but was visible also in the course of the urbanisation processes in the countries of the North. Thus in spite of *prima facie* good reasons to decentralise resources and powers from central to regional and local governments, until very recently, national policies in this direction have remained dead letters (Rondinelli et al, 1984). In the other direction, local authorities, jealous of the small resources and powers that they do possess, have been unwilling to share these with any higher authority and colluded with national governments that are all too happy to keep authorities small and hence minimise any raising of local voices that become evident when cities become politically powerful.

It is perfectly clear that a number of environmental management functions must in virtually all metropolitan areas have a significant regional dimension if they are to function effectively: urban water supply requires management across a wider region and similarly in most cities there must be a regional dimension to flood control. The removal of waste water and solid waste also cannot be undertaken by local efforts alone but requires a regional management function if they are to operate satisfactorily. Transport systems also require a metropolitan dimension to their management - and in particular in the provision of public transport systems - if the cities are to be relieved of congestion. Indeed, it is not difficult to argue that it is the absence of effective regional environmental planning, management and coordination which is the primary reason for the inadequacy of environmental services in very many urban regions of the South.

As cities expanded in the countries of the North in the 19th and early 20th century, these problems were solved in a variety of ways. Local authorities pioneered the production and distribution of gas and electricity and became entrepreneurial in order to buy in water resources sometimes from considerable distances.[4] Transport undertakings developed largely by private companies collapsed as a result of over-competition and the better systems today are the result of coordinated public action. Waste management for sewage or solid waste was never viable simply as a private business: it always required the public establishment of the system as a whole even if in recent years private companies have taken over and improved parts of the function but within the given system and with strong public supervision.

But what is of considerable importance is that it was public pressure in the form of movements within civil society that lay behind the creation of environmental management systems in the cities of the North. The Progressive Movement in the United States and the Public Health movement in the UK, together with a variety of similar movements in the countries of Continental Europe, exerted the pressures and sense of public responsibility necessary to call forth the creativity that produced the management systems and technologies which give us the livable cities we know in the countries of the North today.

It has become evident during this century that it is not possible simply to transfer these methods as methods to the newly urbanising countries. It is necessary for the culture of good governance to arise in the context of each national or local situation. Local civic movements, given coherence by local NGOs and community groups, appear to be a prerequisite to create the context for reform. This in turn forces greater accountability and the imagination to invent locally appropriate technologies and systems of management. This may be able to make use of systems created in the cities of the North or even systems created with assistance of external agencies. But they must be seen to be the product of local will if they are to function and to solve local environmental problems effectively.

Of course the countries of the South have not been free agents to choose their own political paths and for much of the middle years of this century, even after the demise of colonial rule, were locked into the ideological battles of the North: to conform to rules of either capitalism or communism and in either case suffer repressive regimes allowing little by way of local political creativity. With the dissolution of this battle has come another form of ideological straightjacket in a form of 'globalisation' that presumes the pre-eminence of private capital to determine how the world, including cities, should be organised.

This context nevertheless allows for some political debate which has resulted in genuine moves towards reform, including democratisation and decentralisation in most countries of the South. These mechanisms show themselves to be much more complex than may, on the face of it, seem to be the case (Souza) and each new approach seems to throw up at least as many problems as it solves. Of course the proof that reform has genuinely taken root is where problems arising from initial steps towards reform are taken as a starting point for yet new initiatives and not simply used as proof that reform cannot work.

The reform process starts at both ends. On the one hand local community and NGO activity is one of the symptoms of reform and an important component of the eventual solution. There is very widespread interest in reforming solid waste management at the community level to show immediate improvements in the local environment. However, this soon reveals that genuine improvement in urban and metropolitan areas cannot be expected unless there is also reform at higher levels (Kironde; Ratnayake; Cooper).

At the level of environmental improvements particularly, though not exclusively, in low-income housing areas there is already a significant history of modes of internal organisation with government and international agency involvement. Initially very narrow and non-participatory approaches to urban upgrading showed that large investments could rapidly deteriorate for lack of maintenance or lack of a sense of ownership of the results (León Balza). Newer approaches, involving new social forces and involving inhabitants in the process of planning and supervising the work promise significantly better results (Kyessi). But this is still far from contributing meaningfully to broader goals of sustainable urban development (Edelman et al).

In fact much of the reform movement starts simply from growing awareness that approaches that governments have hitherto been taking, and particularly combinations of technocracy and clientelism, are inadequate to confronting the problems faced by rapidly growing cities (Yonder; Mora; Girardin and Greco). But the most promising efforts are those that start at the level of the city and work in both directions, attempting to introduce new decision-making processes that are inherently more democratic and that involve the wider public both in deciding what are genuine problems and only then deciding on appropriate ways to solve them, as opposed to adopting ready-made solutions and trying to fit them to existing conditions.

A start has been made on such comprehensive, holistic approaches in many countries and individual cities around the world. These may be initiated through local coalitions of interest (Biswas) and certain international agencies have been open to assisting to bring these to life (Doe and Tetteh; Atkinson). It is significant, however, that although these point in the direction of sustainable urban development - and use the term in a forthright manner - there still seems to be some way to go before a coherent approach is taken to this question (Allen; Velásquez and Pacheco). In particular few attempts are yet visible to reinvigorate management at the metropolitan regional level with the concepts and strategies of participatory planning and management with sustainability as a central objective (UNCHS/UNEP, 1997).

It might be surmised from the foregoing that, given the obvious differences in the quality of urban life, the current urban management reform movements of the South are not relevant to the North. However, the new environmental agenda, concerned as it is with sustainable development, is in reality particularly relevant to the North which is, after all, dedicated to lifestyles and a mode of organisation that is, in the view of the World Commission on Environment and Development (WCED, 1988) clearly unsustainable.

And this situation has, indeed inspired a whole new wave of social movements and related political and organisational initiatives, particularly under the title of Local Agenda 21. Whilst this title originates in the agreements of the United Nations Conference on Environment and

Development (UNCED), the movements precede this and the initiatives in different countries and cities have their own dynamics and approaches. Between the covers of this book, these are reflected in just two views of the situation in the UK; however, there is now a considerable local literature documenting the rapid growth of these initiatives in the countries of the North, to which interested readers might turn.[5]

Structure of the Book

As noted above, the focus of attention of the papers that make up this book is more upon the city than upon the metropolitan area or the level of the neighbourhood or community. The papers are grouped into three sections: policy, management, and organisation and politics. A paper appears in one of these sections because it had something interesting to say about the group subject. By this criterion, most could have easily claimed a place in any of the sections, so they should all be given attention. It is their collective message which is most unique about them: the breadth of the picture they present of current environmental management in urban areas.

Concerning the development of **policy** aimed at improving urban environmental management, the papers help to identify key issues presently under discussion. These include on the one hand approaches to sectoral problems such as waste management, but on the other hand there is an underlying issue of the understanding of the seriousness of different problems in the overall context of environmental management policy and how priorities should be established. In many contexts the right questions have yet to be formulated; an adequate perception of what might be a rapidly changing reality is lacking. But while research ought to inform policy reform, this is by no means the only missing factor. An adequate consideration of relations among tiers of government, the role of judicial power and the need to guarantee better conditions of access to courts for the defence of collective interests, and a more explicit recognition of the role of stakeholders like NGOs, CBOs and unions in the political process, increasingly provide the background to policy shifts.

Management of the urban environment can be interpreted as simply an extension or adjustment of urban management in general. Indeed, it is apparent from many of the papers in this volume that many urban managers are seeing the problem in this light. However, there are two objections to this interpretation. Firstly the deterioration of the urban environment visible in most cities in the countries of the South would indicate that new approaches to urban environmental management are urgently needed.

The second objection arises from the more coherent approaches, indicated in a few papers in this book (Doe and Tetteh; Atkinson), which are attempting to look at environmental issues in a more holistic manner and at the same time employ methods that involve actors other than local government. Whilst such new approaches are clearly making only slow progress, it does seem that they hold the key to changes in management practices that will eventually come to address adequately the environmental problems faced by metropolitan areas.

Regarding the issues of **organisation and politics** the papers in this volume are useful in identifying some of the underlying problems encountered in attempting new approaches to urban environmental management. Whilst adequate policies might be adopted and appropriate changes made to the approach to managing the urban environment, the difficulties encountered in implementation are many. These arise from problems of the fit between approach and context in institutional terms and then from problems of interests that run counter to changes in the way that things are done. It is at this level that many of the most interesting lessons can be learned that will enable those keen to improve on metropolitan environmental management to confront and incorporate existing interests into new and better ways of doing things.

Notes

1 'Development that meets the needs of the present without compromising the ability of future generations to meet their own needs' (WCED, 1988), which is a concern both for economical use of resources and for the maintenance of the environment.

2 Names of authors in brackets in the text without dates refer to papers in this volume.

3 There have, however, been exceptions, such as in China, where large urban regions were created as administrative areas with a specific policy of maximising self-reliance in food and certain manufactured goods.

4 The New York water utility built a system at the end of the 19th century which brought sufficient water from the distant Adirondak mountains to supply a vastly expanded city. The case of Los Angeles, plundering water resources and destroying the water systems of distant areas, is a well known story (Reisner,

1990). Late 19th century plans to supply London's water from Wales were never realised, but Manchester obtains its supply from the distant Lake District.
5   See Lafferty and Eckerberg (1998) for an overview.

References

Gilbert, R., Stevenson, D., Girardet, H. and Stren, R. (1996), *Making Cities Work: The Role of Local Authorities in the Urban Environment*, Earthscan, London.

Hardoy, J. E., Mitlin, D. and Satterthwaite, D. (1992), *Environmental Problems in Third World Cities*, Earthscan, London.

Haughton, G. and Hunter, C. (1994), *Sustainable Cities*, Regional Studies Association, London.

Lafferty, W. and Eckerberg, K. (eds) (1998), *From the Earth Summit to Local Agenda 21: Working towards Sustainable Development*, Earthscan, London.

Pugh, C. (ed) (1996), *Sustainability, the Environment and Urbanization*, Earthscan, London.

Reisner, M. (1990), *Cadillac Desert: The American West and its Disappearing Water*, Secker & Warburg, London.

Rondinelli, D. A., Nellis, J. R. and Cheema, G S. (1984), 'Decentralisation in developing countries: A review of recent experience', *World Bank Staff Working Paper no. 581*, World Bank, Washington DC.

Satterthwaite, D. (1997), 'Sustainable cities or cities that contribute to sustainable development?', *Urban Studies*, vol. 34, no. 10, pp. 1667-1691.

UNCHS/UNEP (1997), *Implementing the Urban Environmental Agenda*, United Nations Centre for Human Settlements and United Nations Environmental Programme, Nairobi.

WCED, (1988), *Our Common Future*, Oxford University Press, Oxford.

White, R. (1992), *Urban Environmental Management: Environmental Change and Urban Design*, Wiley, Chichester.

PART I
POLICY

2 Policy and Politics in Urban Environmental Management

EDÉSIO FERNANDES

Introduction

Policy, management and organisation are, or should be, inseparable aspects of urban environmental research. Being such intertwined dimensions of state action, they should be considered as a whole by public administrators and legislators, service providers and programme-makers. If anything, policies - materialised in governmental programmes, projects and laws - deserve to be given special attention as they should provide the political, legal, institutional and financial framework within which urban environmental management can take place.

This section explores the importance, nature and conditions of policy-making for urban environmental management. It comprises a broad-ranging collection of empirical studies, some of which are more general than others. They touch on issues as diverse as water supply in Madras, India (Anand); waste management in Dar es Salaam, Tanzania (Kironde); the creation of green spaces in Santiago, Chile (León Balza); the formulation of environmental planning in Cape Town, South Africa (Younge); the definition of an action plan for Calcutta, India (Biswas); and health conditions in urban areas (Stephens). However, at least one major theme links these different chapters: perhaps in a more explicit way than the other chapters in this book, the chapters grouped in this section emphasise the centrality of policies.

Despite the authors' different backgrounds and working realities, several common policy matters and concerns can be identified in their studies. This chapter aims to provide a brief overview of the chapters in this section, and I shall do so by identifying and discussing such common matters and concerns, as well as by making general comments on the question of policy-making within the context of urban environmental management. Special emphasis will be placed on the discussion of the relationship between policy-making and the broader political process.

13

Policy Matters and Main Concerns

The main policy matters discussed in this section include, among others, the nature of the conceptual framework underlying policy-making; the dynamics of intergovernmental relations; the importance of constitutional and legal distribution of powers; and the conditions for administrative efficiency. Common concerns include the need for researchers to have a broader approach to urban environmental realities, as well as for policy-makers to give a comprehensive and integrated treatment to the matter. Most authors especially highlight, or at least allude to, the intimate relationship between urban environmental policy-making and implementation, and the overall political process.

I shall discuss very briefly some of such matters and concerns, as well as make a few comments on the shortcomings of the political analysis underlying the studies in this section.

The Nature of the Conceptual Framework

One of the main matters raised by several authors in this section refers to the need for policy-making to be undertaken within a proper framework, in which, together with an adequate recognition of the problems to be tackled, concepts and principles are clearly defined. Stephens stresses that the manner how a problem is conceptualised and defined affects policy deeply; by the same token, the failure to view existing problems from such an explicit framework - as well as to fully realise their implications - has been largely responsible for the failure of attempts to implement urban environmental policies.

It seems unequivocal that, when not followed by the necessary instruments - technical, financial, legal and administrative - public policies are doomed to failure and generate administrative inefficiency, open room for corruption, and provoke, among many other harmful effects, the waste of already scarce resources and the unnecessary duplication of efforts. They also risk becoming mere rhetorical stances, although it should be said that, even as such, they do play an important ideological role in the state's attempts to legitimise its action vis-à-vis urban society.

The Dynamics of Intergovernmental Relations

Another matter frequently discussed by the authors in this section refers to the fundamental role played by intergovernmental relations in the (un)successful formulation and implementation of urban environmental policies. Many studies suggest that, far from cooperating with and complementing each other, the relations between the several levels of government, particularly between central government and local government, have often been strained and conflicting. The lack of proper coordination and integration, among other factors, has also frequently resulted in the incoherence of many existing policies, the lack of continuity of promising experiences and the failure to improve and replicate relatively successful experiences.

Among the many ramifications of this matter, two aspects deserve special mention. First, it is important to qualify the widespread claim that, per se, the strengthening of local government is the principal way to promote democratisation as well as rational and efficient management. In many countries, a movement favouring localism/municipalism at all cost has emerged as a reaction to a long tradition of authoritarian and centralised governments, in some cases calling for the devolution of political and financial resources to local administrations - while not accepting responsibility for the obligations traditionally associated with central government.

However, without a proper political-historical basis, mere decentralisation does not imply rational management; indeed, as Younge's study of South Africa argues, sometimes in order to decentralise it is necessary to regroup pre-existing local administrations. Moreover, as León Balza argues in the case of Chile, the role of regional and central government in urban environmental management should be redefined, and not ignored as has frequently happened.

Second, in many developing countries one of the most important characteristics of the process of urbanisation has been the formation of metropolitan areas, which are major elements in their overall urban-territorial orders, as well as in their socio-economic structures. However, this fundamental metropolitan dimension has been missing from most urban environmental policies, which have been formulated at the central or the local level. Metropolitan problems cannot be solved by local policies alone or by the combination of isolated policies, requiring instead the creation of specific legal-institutional, if not political, structures. In so far as metropolitan problems are concerned, the true local dimension, that closest to the population, is the metropolitan one. South Africa has been one of the few developing countries that have attempted to face up to this matter, but Younge shows how difficult a process it has been, especially given the way it directly challenges the country's former socio-economic, as well as political, elitist and segregated order.

The Importance of the Constitutional and Legal Distribution of Powers

The discussion above on the need for improving the political-institutional context of intergovernmental relations is directly related to another central, though long neglected matter, that of the need for a proper constitutional and legal framework to be established in order to guide the process of policy-making in urban environmental matters. In fact, the legal dimension of this process - in all its implications - has been largely ignored, when not otherwise taken for granted nor properly questioned. Younge shows the importance of the constitutional-legal framework, as it directly affects the conditions of distribution of political powers, administrative responsibilities and financial resources.

It should be stressed that, in many developing countries, the prevailing legal-institutional order does not properly translate or express the existing urban-territorial structure created by the process of intensive urbanisation. Traditional legal boundaries of municipalities very often do not coincide with the actual urban centres, especially in the case of metropolitan areas. The creation of new and flexible legal instruments to enable state action has been constrained by both conservative legal ideologies and the lack of understanding of the new realities - social, economic, political and legal - brought about by the process of intensive urbanisation.

In this context, the discussion on the conditions of policy-making should be linked to a critical evaluation of the overall legal order and corresponding institutional apparatus prevailing in the country. In the same way that many researchers tend to think of the law as a theoretically un-complicated field of knowledge, policy-makers often see it as a mere instrument for state action, suggesting that urban environmental problems can be solved by the enactment of specific laws. However, as the efficacy of specific laws is determined by the broader context of the country's constitutional-legal order, the successful formulation and implementation of policy proposals very often depend on the undertaking of structural legal reform.

The Conditions for Administrative Efficiency

Many of the studies in this section highlight the problems and obstacles to the formulation and implementation of policies resulting from the maintenance of obsolete administrative apparatuses, which are frequently incapable of responding to existing demands and thus distort and immobilise state action. Kironde shows the risks of irrational bureaucracy, especially concerning the scope it gives to corruption and lack of accountability. León Balza also stresses the need for investment in capacity building especially in local administrations.

Interdisciplinary Research

Urban environmental researchers seem to agree that the proper understanding of the complex phenomenon of urbanisation requires an interdisciplinary approach, which can lead to the constitution of a common language. Moreover, it is widely believed that the formulation of any comprehensive policies on such matters would also require an interdisciplinary approach to the dynamics of urban environmental processes.

The fact is that significant steps have already been taken in this respect internationally, although Stephens argues that researchers and policy-makers alike have neglected the consideration of fundamental urban health issues, thus forgetting the very origins of urban studies and policy-making. I would add that there is also an urgent need to understand the forms, and role, of (il)legality in urban areas (see Fernandes and Varley, 1998). If, as argued above, the prevailing legal order directly affects the conditions of policy making and overall urban environmental management, researchers should also consider the diverse social practices which constitute the widespread phenomenon of "legal pluralism" resulting from the exclusionary nature of the official legal system in many developing countries. Not only do such practices have an impact on the official decision-making and management process in urban areas, but they can also suggest alternative and more appropriate ways of dealing with specific realities.

Multi-sectoral Public Policies

A corresponding concern to the above is that, in institutional terms, the formulation of comprehensive policies also requires a multi-sectoral approach. If intergovernmental relations have been problematic, institutional relations and communication within the same governmental level have not been adequate either; the fragmented, incomplete character of urban environmental policies is often referred to throughout this book. In fact, they seem to remain marginal even within the overall context of public policies.

The integration of the so-called green and brown agendas, as well as the rejection of the traditional environment vs. development stance, is long overdue. Biswas suggests that only an environment-cum-economy approach can lead to the formulation of comprehensive and effective policies. The problem remains that, combined with the inadequate distribution of political and legal powers and of financial resources, not only have sectoral policies failed to express the common language of interdisciplinary research, but, as Stephens argues, they have also tended to obey the entrepreneurial logic of the economic groups and interests which have controlled much of the state apparatus in developing countries.

Politics Beyond "Political Will"

The studies in this section indicate that the success and/or failure of urban environmental policies depend on several inter-related factors, some of which were discussed above. However, it could be said that most of them, as for that matter many critical urban environmental studies, ultimately seem to credit the failure or success of official policies to the existence, or absence, of "political will". Such a vague, undefined concept seems to refer to some sort of effective governmental commitment to the implementation of programmes and to the enforcement of laws which materialise public policies in the field of urban environmental management.

It is not clear from such studies who, if anyone, has "political will", or why it frequently does not exist in so far as urban environmental policies are concerned. So, for example, whereas the lack of "political will" is largely to be blamed for the failure of disaster prevention programmes in

Central America (Mora, 1997) and the shortcomings of the waste management programme in Dar es Salaam, the existence of "political will" has made possible the relative success of the green parks programme in Santiago.

I would argue that, especially considering the advanced stage of urban environmental research, it is no longer sufficient to have recourse to such a generic - at once all-encompassing and hollow - explanation without qualifying it further. The fact is that, regardless of their increasing interdisciplinary approach, most studies in the urban environmental field still reveal a lack of understanding of the political dynamics and realities which underlie and structure the process of intensive urbanisation and environmental degradation, as well as determining the conditions of policy-making, programme-implementation and law-enforcement.

Several studies have highlighted the direct links between poverty and the process of environmental degradation in urban areas, but they still lack an understanding of the political-economic processes which have determined both phenomena. By the same token, little has been written on the nature of the political interests which have marginalised urban environmental matters and policies, if not completely excluded them from the official decision-making process of the public order. Nor do we see analyses of why certain policies obtain political support.

In the absence of a deeper and comprehensive political analysis, the formulation of policy proposals is very often (mis)guided by unrealistic, though well-intended, concepts and principles, or by technocratic and pseudo-rational analyses and justifications. Policy-making is reduced to wishful thinking.

By the same token, many policy evaluations tend to adopt a pessimistic, if not nihilistic tone, which can be noted in many of the chapters in this book. Given the extent of social, urban and environmental problems, especially in developing countries, many analysts seem to succumb to the bleakest prospects. At a loss to explain properly the failure of policies and programmes, they also tend to, in the process, dismiss the actual advances and achievements already made in several countries. The failure to understand the configuration of political and historical forces which has made such experiences possible means, among other things, that they cannot be replicated.

Indeed, even given due consideration to all the existing problems and obstacles to urban environmental management, important experiences have been successfully undertaken, especially at the local level, and the principles underlying many of them could be possibly transported to other realities (Fernandes, 1998). This is not to suggest that a proper consideration of the political process would necessarily point towards solutions to the existing problems, but at least it should help to focus the discussion and channel the incipient resources available.

Few studies reveal an objective understanding of the fiscal-monetary realities prevailing in developing countries, and of their inseparable relation with the political process (see Souza, 1997). It is imperative that the matter of public finances be confronted in a more sensible way by policy-makers. Moreover, the incipient political analysis underlying urban environmental studies have largely concentrated on traditional relations between state and society, when in the last two decades substantial changes have taken place in the nature and role of the state, in the state's relations with the market and society, as well as in societal organisation. As yet, such changes have neither been fully acknowledged or properly understood. León Balza stresses the importance of the involvement of the private sector in policy-making and partnership schemes. Kironde argues that informal, active agents have not been taken into account by policy-makers. Anand argues that the failure to understand the market often leads to increasing long standing inequalities in the provision of services.

Combining Representative and Participatory Democracy

At this point, I would suggest that, if decentralisation does not necessarily entail democratisation, as argued above, representative democracy does not encompass the whole political process. Especially in developing countries, traditional institutions of democratic representation - exclusively based as they are on the action of restricted elected political-institutional bodies - no longer express their overall lively and contradictory political process. The official apparatus in force has failed to recognise the actual conditions of political mobilisation in urban areas, which involve the participation of several new socio-political actors such as urban and environmental NGOs, CBOs, residents associations, and trade unions.

If, as argued above, the intricacies of the dynamic informal economy and the many forms of legal pluralism should be acknowledged by policy-makers and legislators, including during the formulation of privatisation schemes, also the new forms of societal organisation must be recognised and incorporated into the decision-making process of the public order. Yet, Younge's study of South Africa's recent experience - always emblematic of the possibilities of a renewed socio-political organisation - suggests how, when it comes to the definition of a politico-institutional treatment of a given reality, conservative solutions have been given priority, to the detriment of a more lively and participatory process. The fact that traditional political institutions are exclusionary of society cannot remain unchallenged.

Many of the studies in this book emphasise the need for increased popular participation in the process of policy-making and urban environmental management. I would argue that only a combination of renewed representative institutions and new forms of direct participation, including NGOs and other forms of collective interest aggregation, can effectively lead to the formulation of technically adequate, socially just and politically democratic urban environmental policies. An interesting example of such a combination, with direct implications for policy-making, is that of the "participatory budget" which has been adopted by several Brazilian municipalities for the last ten years or so, in which the local populations, through their social and collective organisations, have effectively participated in the determination of the priorities and conditions of the allocation of part of the local investment budgets (Fernandes, 1996).

As I have stressed, a fundamental political issue to be confronted concerns the role of law-making and enforcement for policy-making. As the new attempts at urban environmental management are taking place within the broader context of political re-democratisation in many developing countries, the importance of law reform and judicial review cannot be underestimated. In particular, as Biswas argues, the role of judicial power and the need to guarantee better conditions of access to courts for the defence of collective interests - expressed in official policies, programmes and laws - are crucial matters in that they open more assertive scope for actions by urban and environmental organisations (see also Fernandes, 1995).

Conclusion

As a whole, the studies in this section are indicative of the advances and shortcomings of both urban environmental research and policy-making in developing countries over the last two decades. They provide grounds for both hope and despair, but above all they suggest that there is a vast room for improvement in such inter-related fields. It could be said that, to a significant extent, the failure to find adequate answers and solutions for urban environmental questions and problems has been due to the fact that the right questions have frequently not been formulated.

I would argue that only through a better understanding of urban politics - considered in a broad sense, including its legal dimension - can researchers and policy-makers come to ask and answer such questions, formulate and enforce more consistent policies and, hopefully, contribute to diverting the course of the existing processes of urban environmental injustice and degradation.

In this context, a fundamental discussion which cannot be avoided any longer by researchers and policy-makers in many countries refers to the implications of the powerful combination between the concentrated land structure, the prohibitive land market (and the corresponding informal means of access to urban land); and the widespread recognition of virtually unlimited individual property rights, which lie at the root of the segregated urban structure, of the environmental degradation process as well as of the exclusionary political system. However, given the fact that the solution to many urban and environmental problems depends both on more effective state action and on changing the approach to traditional, individualistic property rights, the paradox to be faced by social, urban and environmental movements is: how can such proposals be reconciled with widespread liberalisation and privatisation policies?

It seems that, especially in the current context of redefined state action, the final balance will depend on the existence of solid social institutions and a consolidated public sphere. Education and information policies such as appear in many of the cases collected in this book are of utmost importance in this process, but, in the last analysis, accountability and transparency of the decision-making process will depend on the level and quality of popular participation involved.

References

Fernandes, E. (1996), 'Participatory budget: A new experience of democratic administration in Belo Horizonte, Brazil', *Report*, vol. II, pp. 23-24.

Fernandes, E. (1995), 'Collective interests in Brazilian Environmental Law', in D. Robinson and J. Dunkley (eds), *Public Interest Perspectives in Environmental Law*, Wiley, Chichester, pp. 117-134.

Fernandes, E. (ed) (1988), *Environmental Strategies for Sustainable Development in Urban Areas: Lessons from Africa and Latin America*, Ashgate, Aldershot.

Fernandes, E. and Varley, A. (eds) (1998), *Illegal Cities: Law and Urban Change in Developing Countries*, Zed Books Ltd, London and New York.

Mora, C. S. (1997), 'The promotion of disaster prevention in urban areas into the development agenda and as a state policy in Central America: Social, political, and economic factors', paper presented at the International Conference 'The Challenge of Environmental Management in Metropolitan Areas', organised jointly by the Institute of Commonwealth Studies and the Development Planning Unit of the University of London, 19-20 June.

Souza, C. (1997), *Constitutional Engineering in Brazil: The Politics of Federalism and Decentralization*, Macmillan, London and St. Martin's Press, New York.

3 Creating Metropolitan Environmental Strategies

KALYAN BISWAS

In this chapter, I shall discuss how metropolitan development issues have undergone a significant change in planning and operational terms over the last three decades. Most of my observations and conclusions are drawn from a large number of metropolitan development programs in developing countries, though no specific reference will be made explicit, excepting a few from the Calcutta experience in which I have worked extensively. A vast literature exists on the subject of urban environmental management, and liberal help has been taken from it as appropriate. Emphasis has been given to policy and program issues, rather than to technical and methodological aspects.

Until the late 1980s, it was not fashionable to talk about urban or metropolitan environmental programs, and urban infrastructure development was identified and pursued by all as a major area of planning and investment. The preparatory years for the 1992 Rio de Janeiro Conference brought the environmental discussion to the forefront in a widespread manner. Thereafter, it has become imperative to talk and write about urban environment only, almost as if urban infrastructure development is no longer *de rigueur*. This transition should not be explained as a mere terminological preference or as one of those paradigmatic switches socio-economic planners usually indulge in. There are a good number of reasons why the conventional urban infrastructure development programs had to be reoriented to address environmental issues in a more systematic and holistic manner:

1. Infrastructure is a necessary, but not sufficient, condition to ensure sustainable urban development. Mere improvement in the supply and availability of civic services does not, per se, decrease the overall vulnerability of citizens from environmental hazards. If infrastructure alone were to make a good environment, all the new towns in developing countries created at enormous public

25

expenditure should be considered environmentally sound.

2. Urban infrastructure investment has been largely targeted at residential/area development and has not always been seen as an essential input for industry, trade and services, for the productivity of the urban economy or to mitigate pollution.

3. There are many environmental outcomes of urban activities like industry, transport and energy which have not been taken care of in conventional urban development programs. From the environmental angle, these outcomes need special attention. More water to drink does not decrease non-water borne related mortality or morbidity. Poverty does not increase or decrease because of low-cost sanitation.

4. While many infrastructure projects have succeeded in physical terms, they have not achieved city or sector wide policy or institutional impact. Removal or easing of traffic bottlenecks, or creation of new bypasses, for instance, are now perceived as incremental short-term solutions to complex urban problems requiring a comprehensive urban transport strategy linked to land planning, investment, resource recovery, pricing and environmental concerns.

5. Urban infrastructure improvement programs have been seen as engineering solutions to growing demands. They have been more often characterised as technocratic interventions to provide supply side solutions to demand estimates. Environmental dimensions have not usually figured in the assessment of the needs or in the analysis of the consequences of investments.

6. Urban development programs have so far mostly been project bases, not policy based.

Agency Determines Function

It should be stressed that many urban development programs deserve praise. In key areas like water supply, sewerage and drainage, traffic and transportation, slum improvement, housing and new area development, significant advances have been made. In clearing backlogs and creating favourable situations for future expansions, such investments have certainly turned the tide against the downward slide in urban living. In those metropolitan areas where special or para-statal agencies have been created to plan, coordinate, implement and monitor multi-sectoral investment programs, much has been achieved in terms of professionalism, training, surveys, competence, tighter program budgeting and administration, enactment of new legislation and regulations, and a level of credibility and public image.

Yet, analysts have not failed to point out certain lessons which need to be learnt from the decades of infrastructure development programs, especially concerning their planning and institutional aspects:

1. Top-down approaches can remove some major bottlenecks in system functions, but they do not result in sustainable street level and area wise environmental improvements.

2. Large centralised top-down programs require large centralised top-down institutions, which tend to be inward looking, concerned with their own survival and predominance, thus often bypassing the larger interests of the community.

3. Large centralised institutions need large projects to sustain themselves, in which process they often develop a "public works" culture of their own at the expense of necessary software support.

4. Projects planned and executed by large centralised institutions often exclude smaller and lower popular organisations from the process of consultation and participation, thereby imposing solutions on communities without reference to their long-term needs.

In such cases, institutional arrangements for implementing development projects usually take the form of the development of the main project institution responsible for centralised planning, budgeting, supervision and coordination. The external relationships of such institutions do not get emphasised, as a result of which the web of institutions responsible for environmental management does not get much attention. New legislation is resorted to as it is not a costly proposition, but it is often deemed to be expression of a political will. Difficult decisions concerning fiscal reforms, resource mobilisation, community involvement and process improvement involving networking often do not get done or receive lesser attention. The problem is aggravated when there is a resource crunch and the institution is no longer able to leverage additional national or international assistance.

Some questions have remained unanswered through the history of urban metropolitan development in developing countries, namely: has the urban development that has taken place degraded or upgraded the quality of air, water and land? Has it been environmentally sensitive (resource depletion and non-replenishment), and has it boosted the urban economy? Has it aroused involvement of public agencies, private business, NGOs and CBOs? Has it led to better health conditions, especially less mortality and morbidity?

Environment-cum-Economy Approach

All cities are costly. Bigger cities are costlier. The cost of a city and its environmental outcomes are dependent on city structure, lay out and form. Compact city, high density, activity intensity, transit system, etc., determine the cost and spread of services vis-à-vis dispersed settlements. Quite often urban demand for food, water, minerals, fuelwood, fossil fuels and other resources affect distant populations, forests, and watersheds; costs increase. On these features - as also on how the city and civil society function - depend the energy efficiency, consumption pattern, resource use intensity, etc. How the economy and the activities of society impact on the environment is as important as how the environment itself influences the economy and society.

It is now being increasingly acknowledged that cities play important economic, social and cultural roles. A bad environment makes a city an unattractive place to live, work and invest in. The development potential of cities is threatened by environmental degradation, and unless efforts are made to improve it, the economic growth of cities may suffer. Efficient and productive cities are essential for national economic growth and welfare; equally, strong urban economies generate the resources needed for public and private investments in infrastructure, education, health and improved living conditions.

An urban economy may be under threat because of the pressures placed on the environment by that economy. An economy under threat is as important as an environment under threat. However, sound environmental management and improved wealth generation are not necessarily mutually exclusive. What is required is an economic-cum-environmental management strategy. Environmental changes require difficult political and economic trade-offs, the most typical ones being made when economic and environmental objectives conflict. These trade-offs occur at several levels, namely: between the increasing productivity of cities due to economy of scale and agglomeration and the increasing cost of providing environmental infrastructure and services; between different strategies and policy instruments for achieving effective environment management; and between political motives and economic expediency.

As nations and cities aspire to develop, restructure and compete in the international market through deregulation, liberalisation and export oriented growth, there is likely to be more environmental degradation in the short- and medium-term, before capacity increases to match and control the damages. There is bound to be a trade-off between industrialisation, economic development and environmental protection. If in the medium-term resource management and capacity building do not take place as fast as necessary (a characteristic of developing countries), a new kind of environmental management strategy and action plan is called for.

Metropolitan Policy and Development

If the existing urban literature has long criticised the physical determinism approach towards city planning and development, the current discussion is taking place in altogether transformed global and national economic contexts. Macroeconomic stabilisation with deregulation and a greater recognition of market forces seem to be in the order of the day in most countries. In this context, the development of policy instruments for environmental improvement has been limited by several factors, including the fact that, in contemporary industrialisation, urban specificity is becoming less and less clear. There are a large number of public activities that are increasingly influencing, directly or indirectly, the general process of urbanisation, and it is not always easy to isolate urban activities which influence or are intended to influence the affairs of at least the major urban areas.

Basically, there is a problem of identifying what policy aspects to exclude from urban policy, the locational impact of any major expenditure, and the impacts of "unviable" policies. In other words, urban policy is already being made in a surrogate manner under some other name. It is not clear to what extent environmental degradation is a necessary or inescapable result of urbanisation and economic change. Environment is deeply rooted in the way in which both civil society and the economy are structured and functioned. Changing the way they work is complex, in as much a political and social process as a technical one, and takes time. If urbanisation is one of the methods by which these changes take place, then its repercussions or benefits on the environment will be determined accordingly.

Metropolitan environmental improvements are better understood in city-specific terms, but there is an important role for national governments in addressing these problems through national strategies and action. Where urban environmental issues are identified as a national priority, they usually do not appear to be based on an in-depth analysis of urbanisation trends in the country or the nature, extent, causes, and national significance of urban environmental problems. An urban management strategy needs to be developed in coordination with plans for urban infrastructure development, for economic development, energy development, settlement locations, water resource management, etc. Where these other plans do not exist or are incomplete or outdated, the environmental management strategy preparation is more difficult, but can provide the opportunity to initiate or update planning in other related areas.

Ideally, metropolitan environmental management strategy should be viewed as a consensus document by local authorities, community groups, central and regional sector agencies, the business community and others. There has to be a custodian of this strategy, whose responsibility is to get it implemented; in the absence of any better expression, let this custodian be the ubiquitous "environmental governance". There are various levels of this governance whose attributes *inter alia* would be capacity for planning and implementation; financial management of urban bodies; availability and deployment of trained manpower; and political will and determination for implementation.

Positioning the CEMSAP

At this stage, I would like to highlight the work that has been done in the preparation of Calcutta's Environmental Management Strategy and Action Plan (CEMSAP) with technical assistance from the then British Government's Overseas Development Agency (ODA), now Department for International Development (DfID). Over the last two decades and a half, Calcutta's urban development programs with World Bank and other assistance have focused on augmenting and strengthening basic infrastructure services. As a result, environmental conditions are significantly better today than what they might otherwise have been. But severe pollution and environmental degradation problems remain, since these were not dealt with as part of the urban development programs. The economic losses in terms of health, productivity and amenity costs due to current levels of environmental degradation are very large. Increasing levels of pollution represent a significant threat to existing and new private sector investments. Many of Calcutta's poor have not had access to formal education or are not reached by informal training and awareness campaigns. They have a low level awareness about the causes of environmental problems and alternative solutions.

CEMSAP has aimed towards, and largely succeeded in, making environmental management and protection an explicit agenda and action point in the ongoing urban and other development programs in Calcutta. It is now geared to complement urban development programs by adding another dimension to it. It has addressed the environmental opportunities and constraints for achieving the goals of development programs of the government in the metropolitan area. Widespread consultations with primary and secondary stakeholders and user groups have been held to appreciate their perceptions, capacities, willingness and cooperation to devise and implement relevant projects.

CEMSAP is founded on the recognition that poverty and inadequate resources management define the characteristics of Calcutta's environment. It has addressed these issues, going beyond mere improvements in the physical environment. Continued environmental degradation and service deficiencies will not only undermine long-term growth, but will also hit the urban poor even harder. A whole range of issues like infrastructure, productivity, employment, efficiency, health, investment, poverty reduction, wealth generation, etc., are confronting policy-makers in Calcutta. CEMSAP encompasses the economic and social environment of Calcutta as much as the metropolitan physical environment.

Put simply, CEMSAP's vision is to make Calcutta "a better place to live, work and invest in". But this statement has many operational implications, such as to increase the overall competitiveness of the city for investment and business which is also associated with the quality of urban environment; promote environmental justice by way of institutionalising the incentive structure governing resource use and provision of environmental services; and provide the basic minimum level of city services with equity and efficiency.

In many ways, CEMSAP is an attempt to break with the past to create an enabling framework for the implementation of appropriate environmental policies and practices. It aims to improve the environment in order to realise sustainable health and economic benefits. In more specific terms, a distinguishing feature of CEMSAP when compared to similar exercises elsewhere has been the fact that, as it has spent less time on primary surveys and depended on secondary and other published data, it has been able to produce strategy and action plan documents in about 20 months. It continues to emphasise investment in civic infrastructure, but looks at broader issues and has framed policies and programs beyond infrastructure. It recognises that poverty remediation and economic growth should be the central focus; wealth creation, job opportunities and environmental protection should and could be a win-win-win situation.

Moreover, CEMSAP has worked out the priorities, the acceptability and the sustainability of infrastructure and environmental investment on criteria of health, distribution and efficiency. It has advocated community level approach and local capacity building so that these contribute to more meaningful planning and management.

The Challenges of Metropolitan Environmental Management (MEM)

Cities in developing countries are faced with serious pollution problems before controls over traditional pollution sources have been put in place and before strong institutions have been developed. Although developing cities can take advantage of the "appropriateness" of the latest anti-pollution technologies by internalising their costs, such technologies remain costly. In view of the urge of economic development in developing countries, it is likely that for some time in future urban environmental conditions might actually deteriorate. With more production in the existing institutions, diversification, export oriented growth and more intensity in resource use - all of which are the features of accelerated industrial development - urban environmental conditions may first degrade before they improve.

Metropolitan environmental policy formulation and implementation require difficult political and economic trade-offs, and policy reforms of three kinds are necessary, namely: market based policies, which use pricing, taxes, or marketable permits to modify behaviour; regulatory or administrative policies that impose quantitative restriction, enforce property rights and safer investments; and extra regulatory approaches to pollution control, such as environmental audit, public liability insurance, etc.

This leads to the question of environmental governance. A central issue is how the formal bureaucracy in the government reorganises itself to address the demands of environmental management. The Calcutta experience shows that all governmental departments, like environment, transport, industry, urban, municipal affairs, health and a number of agencies, are directly involved in achieving coordinated action and policy planning. While re-engineering of these units is required, good governance also demands their interface with trade unions, Chambers of Commerce, academic and professional institutions, NGOs, CBOs, etc. In fact, an altogether new administrative culture is needed to be developed over time through a process of retraining, consultation, and interface, to a degree not practised before.

Related questions include: do market forces generate concentrations in metro areas and consequent negative externalities? Do we have to rethink rules about economics of scale of agglomeration? Do we have a basis of identifying a body of proven sustainable urban/metro practices, or a mix of measures, in a city?

Reversing the deterioration of urban environment without slowing down economic development requires an environmental strategy that includes a wide range of actions, difficult political and economic choices, and a complex set of natural, social and economic relationships. In many countries (developed and developing) the gap between business and political perspectives is widening. If all politics is indeed local, with increasing demand for decentralisation and empowerment, how can this be matched with the demands of an economy that is on the way to becoming global? Then again, is there any reason to believe that a mix of privatisation and decentralisation actually reduces urban environmental problems? On the other hand, if the role of the government is to be limited, will it affect the execution of environmental plans? Such questions should challenge all nations regardless of their level of economic development or political systems.

It is well known that, among other things, energy and efficient use of land can influence the quality of urban life. The interface between land use, transport and environment is particularly important. Key issues include distortion in urban land markets; ineffective land management policies; and failure to anticipate and properly address growing demand for access. Yet developing cities also have limited financial, technical and institutional resources to tackle such issues.

A closely related challenge is the fact that environmental degradation in metro areas has resulted from inadequate economic development, infrastructure and technology. Economic decline or stagnation coupled with increasing urban concentration have forced an intensive use of an already shrinking natural resource base. Urban industry clusters result in two particular environmental effects, greater vulnerability to industrial hazards and pollution of water and air. Bigger cities in developing countries may be engines of growth, but their potential is not always realised because they are also the sources of concentrated environmental problems. Here the concern is to determine the extent to which environment is both an opportunity for, and a barrier against, economic development.

Underlying such challenges is the need for augmenting managerial and educational resources. This involves a whole set of action areas from development capacity building for planing and implementation, better local financial management and popular participation to technology upgrading. It requires reorienting the training of many professionals, and achieving rewards of trust by better performance. Implementing MEM strategies is even harder than preparing them.

Since political will and determination are indeed the driving forces, professional advice and persuasion should be able to point out clearly the costs of "doing nothing" and to suggest what exactly the short- and medium-term "do-ables" are. In many developing countries, environmental protection and management is still not on the agenda of the political parties, because such issues are still not seen as "votable" and therefore eminently "negotiable". Lip service through rhetorical obeisance is still the order of the day, though the pressure of NGOs and CBOs is mounting, and compliance to norms and standards are reluctantly agreed to. For example, jurisprudence in India allows individuals and community groups to file Public Interest Litigations in the courts for redressing what they perceive to be environmental threats or hazards. Open court hearings are held, and environmental issues are debated in a more transparent manner. Recent changes in Indian laws also make it obligatory to hold public hearings on all developmental projects which require statutory environmental impact assessments.

Yet, it is also possible that environmental improvement programs without regard to social concerns may on occasions increase the plight of the poor. A classic example would be the strict adherence on the "polluter pays principle", which often leads to the closure of industrial units with a downside on the labour front. Removal of street side hawkers and shifting out informal business to the peripheries are other similar measures which attract adverse notice and social unrest when not accompanied by relevant safety nets.

A professional response to convince politicians of taking a pragmatic and responsible approach could be to highlight the political importance of the benefits of MEM in a language which can be easily understood, e.g., by viewing MEM as an economically viable activity (emphasising reusable and renewable resources, and encouraging environment and pollution related investment in industries and services); viewing MEM as a complementary thrust to urban and national economic development; providing yet another opportunity for strengthening the role and capacity of local government; and improving the image to attract national and international investment.

The main political mileage should be extracted from economic benefits. A positive pro-environment image attracts business, tourists, conferences, students, professionals, etc. Improving environmental attraction can also retain indigenous industry and skilled workers. Concentrating on environmental issues helps project the face of a responsible city/national government.

A Suggested Action Plan

Any environmental agenda for developing metropolises has to aim, whenever possible, at a convergence of pro-development and pro-environment goals. Given the absolute base at which many developing countries start their race for development, the pressures of development are likely to have their impact on the environment, before awareness, enforcement, political will, etc., are able to restore the balance. Mitigation and reduction of environmental problems will remain the goal for quite some time.

In such circumstances, the suggested action plan will have to start at the very basics, at a modest level, until it can be scaled up given other inputs. The new agenda would aim to enforce legislation (pollution control, environmental and occupational health, etc.); ensure equitable provision for water supply, waste management, etc.; coordinate land planning, transportation, settlements and environment; and ensure health care provision to treat not only environment related illnesses but also to implement preventive measures.

The new agenda would also aim to evaluate environmental and health impacts of existing production and consumption processes, and take decisions regarding new developments accordingly; implement regulatory and economic investments for internalizing environmental costs into decision framework; ensure coordination and monitoring for effective implementation, review and adjustment of such policy investments on a continuing basis; and increase the capacity of the relevant agencies, local bodies, professional groups, etc., to identify and address city/local level environmental problems. Finally, it would aim to adopt a consensual and participatory approach, relieve economic stagnation, and ensure governmental commitments.

The extent to which environmental quality is achieved in cities is a revealing indicator of the capacity of the concerned agencies. The quality of governance determines the extent to which a city takes advantage of being the centre of concentrated production and population, and avoids or reduces the potential disadvantages.

Conclusion

Many cities are already working in various ways, with a degree of success, towards improved environmental planning and management. Steps have been taken towards better environmental information and expertise, environmental strategies and decision-making, effective implementation, enhanced institutional and participatory capacities, and efficient use of scarce resources. International support programs have been developing, through years of experience, a variety of increasingly effective strategies for cooperating with, and supporting, cities in implementing their urban environmental agenda. There is no better way to achieve success in future.

4 Issues of Inequality in Managing Water Supply in Asian Cities

P. B. ANAND

Introduction

This chapter highlights some aspects of water markets in metropolitan cities with a case of Madras, India. It emphasises the need for developing methods and instruments to understand the distribution and equity aspects of water supply plans and projects. A water balance sheet is constructed for Madras, showing both the sources of water and the destination uses. Based on this and household survey data, the per capita availability of water from different sources is estimated. Distribution of households as per water endowments is examined. Also, some estimates of the differences in average water costs faced by different users are discussed. It is argued that, if the existing endowments are not changed, even a doubling of the total quantity of water will have no impact on the poor. Some aspects of institutional reforms are discussed.

Inequality is an issue of great concern to economists and planners. One of the basic goals of planning is to address inequality, especially so in the context of Asian metropolitan cities. Yet, in the absence of *ex ante* measures of distribution of benefits, many programs, plans and projects are developed with a "notion" of their impact on distribution. For instance, project reports include a column on "beneficiaries" and list that "all the people living in the city (or region) will benefit". Some reports mention that "the poor are more likely to benefit" because the plan or project increases the overall quantity of infrastructure. But how could one be sure that this is indeed the case? For instance, in case of two similar projects, but with different distribution impacts, a generic or subjective assessment would be unable to distinguish between the projects. This issue could be even more serious in case of policies or projects concerning environmental quality. Often, distribution questions are brushed aside or they are confined

37

to theoretical discussions.

The aim of this chapter is to discuss some aspects of inequality in water supply systems and to argue that it can be included in evaluation exercises. Given the impossibility of discussing the full details of the evaluation exercise, this chapter looks mainly at how some simple measures can be developed to understand equality issues in water supply and raise some policy implications.

The study is based on a household survey of 200 households drawn by a multi-stage cluster sampling design. The survey was conducted in the form of structured personal interviews, during June - October 1996. Also, qualitative focus groups, data from depth interviews and participant observation were used to understand water markets in Madras.

Water Supply System in Madras

Accounts of water supply sector predominantly reflect the "formal" or "documented" view of the system, based on information readily available. For instance, the water supply chapters in an earlier study (Dattatri and Anand, 1991), or the draft Master Plan for Madras (MMDA, 1995) are limited to such documented view. Information on water markets is scarce and often is (when available) of dubious basis, though that need not be an excuse. This study is one of the first attempts to collate and understand the different components of water markets in Madras.

Madras, a city of about five million people, has the distinction of being the metropolitan city in India with lowest per capita supply of just about 70 litres per capita per day.[1] In terms of institutional arrangements, four distinct systems of water supply can be found, namely those of the Madras Metropolitan Water Supply and Sewerage Board (in short Metro Water) mainly for Madras City; the nine towns adjoining Madras and forming part of Madras Urban Agglomeration (MUA); self-provision by many households and industries which drill shallow wells or deep tubewells, and the private market, involving both bulk supply by means of tanker trucks of 12,000 litres capacity and the retail distribution of "mineral water" in jerry cans of 10 or 12 litres capacity.

Households sometimes depend on more than one source of water, as is evident from the data in Table 4.1.

Table 4.1   Sources of water for Madras' households, 1996

Source of Water	Madras City	Nine Towns	Total for MMA
1 No source within the premises	19.3%	16.1%	17.6%
2 Shallow well	16.5%	54.8%	29.1%
3 Tubewell	6.4%	--	4.7%
4 Shared municipal tap connection	8.3%	--	6.1%
5 Municipal tap connection	4.6%	6.5%	4.7%
6 Well and connection	15.6%	16.1%	14.9%
7 Tubewell and connection	27.5%	3.2%	20.9%
8 Well, tubewell and connection	1.8%	3.2%	2.0%
Sub-total for households with connection (categories 4+5+6+7+8)	(57.8%)	(29.0%)	(48.6%)
Total for all categories	**100.0%**	**100.0%**	**100.0%**

Source: Madras household survey, 1996.

It can be seen that about 42% of households in Madras City and more than 70% of households in the rest of MMA are not covered by the piped water supply system. The "documented" view in the case of Madras city would say that 92% of the area is covered - and hence that 92% of the population is "covered".

Towards a Water Balance Sheet for Madras

Estimating a water balance sheet for any metropolitan city is difficult, and Madras is no exception. Five sources of information have been used in this exercise, namely: housing tables from the Census 1991, giving information on number of households with the principal source of water; a household survey conducted by me during June-October 1996, giving details of all the sources that a household has; the Annual Report of Metro Water for 1994-95, containing financial statistics; the Madras Metropolitan Development Authority's Draft Second Master Plan; and MMDA-TRF research study reports (mainly Srinivasan, 1991, and Dattatri and Anand, 1991). This estimate relates to the period prior to September 1996, (i.e., before the release of 200 million litres per day or MLD from the Krishna

Water project). The water balance sheet for Madras Urban Agglomeration is presented in Figure 4.1.

The total quantum of water supplied by Metro Water in recent years is 433 MLD (MMDA, 1995). This is drawn from six sources. About 200 MLD comes from the Poondi - Sholavaram - Red Hills system. This is mainly from surface sources, namely the Arniar and Kortaliayar rivers. Another 148 MLD is pumped from a number of well-fields in North and North West Madras. The South Coastal Aquifer provides about 10 MLD. Metro Water has been maintaining some local sources and wells in Urur, Thiruvanmiyur, Porur, Kattupakkam, etc., which together contribute 20 MLD. There are 35 Municipal wells whose potential has been assessed as 5 MLD. These are not connected to the Metro Water distribution system. In addition, there are 7,141 India Mark II pumps and 1,884 tubewells (total: 9,025 MLD). The total yield of these pumps is estimated to be 50 MLD (MMDA Master Plan). Of this, about 71 MLD (i.e., about 16%) is supplied to metered customers. The remaining supply is un-metered.

Figure 4. 1 Water Balance Sheet for Madras Urban Agglomeration, 1996

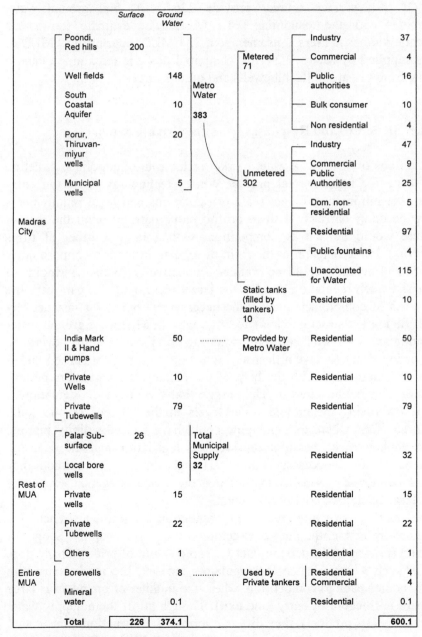

Source: Prepared by the author.

According to the water balance sheet, Madras consumes over 600.1 million litres of water per day (MLD) or roughly 111 litres per capita per day. Of this, surface sources provide 226 MLD and ground water contributes with the remaining 374.1 MLD. The estimate shows that residential/domestic users consume about 324 MLD, but only 189 MLD of this is supplied by Metro Water or Municipalities, the remaining quantity coming from private wells, tubewells and other sources.

Market Structure and Responses from the Private Sector

In conditions of scarcity, a water market reflects an oligopolistic structure with scope for super-normal profits. While technology is not an entry barrier, the minimum efficient scale of operations and legal requirements act as the entry barriers. If there are no entry barriers, then the market structure would be that of competition, with a large number of firms (operators). Panic reaction in the form of regulation tends to impose entry barriers and actually could be counter-productive from the viewpoint of efficiency. Such regulation benefits the large operators (who can incur the legal costs of compliance and get the necessary permits, for instance) by taking the market structure closer to oligopoly. In Madras, a ground water regulation act was introduced eight years ago and the act requires everyone transporting water to have a license. From time to time, unlicensed tanker trucks are impounded with the help of the police. The regulators believe that this is in public interest. They argue that the effect of regulation is seen in stabilising the ground water levels in the villages on the south coastal aquifers, where over-pumping could have caused salinity ingress. All the tanker truck operators consider this legislation and the licensing requirements an unnecessary hurdle. Though that is the opinion they express, economic rationality implies that the business agenda, especially of the large operators, could be different.

The market structure is partly regulation-determined. When the regulators are active, the market structure is close to oligopoly. Scope for collusion (pricing by leadership) and for some extent of price warring does exist in such a case. When the regulators are lax, the market structure quickly comes back to competition, where the number of operators is large and there is freedom of entry (and exit). Though all of them supply water (a homogeneous product), they differ in terms of information, consumers' perceptions about quality of water and the reliability of the operator.

Hence, there is a product differentiation - the product being "the transport and supply of water" and not water *per se*.

There are about 150 tanker trucks operating in the private sector. There are no statistics on concentration ratios, but top five firms have a market share of about 50%. To break even, each truck must make at least 5 to 6 trips per day. On this basis, the total supply by this sector is put at 8 MLD. However, the operators pointed out that they may make even 10 trips during the peak demand period. At least half of the demand is from non-residential users (offices, factories, construction sites, hotels, hospitals, marriage halls, etc.).

With regard to the mineral water markets, the overall size of the market is put at 0.1 MLD, though Madras is one of the fastest growing markets for mineral water consumption.

A third type of response is in the form of rent seeking behaviour for the water supplied through static tanks. These number about 3,396, and each is about 3,000 litres in capacity. They are filled every day by tanker trucks hired by Metro Water from private operators. This is a gratuitous supply for slum dwellers and others not having yard tap connection. However, it was observed that the tanks are regulated by local individuals who collect Rs. 0.25 per kudam (a container of 20 litres capacity) from the consumers.

Distribution Equity

The share of water available to different income groups cannot be worked out without detailed consumption surveys. From the limited data available, I have worked out the average water consumption rates for different users (cf. Table 4.2). For instance, from the Water Balance Sheet we know that un-metered residential consumers get about 97 MLD. According to the 1991 population census, we know that the number of households having a water connection within their premises is 205,765, and the average household size is 5. Therefore, per capita consumption comes to 94 litres per capita per day. Another example can be of those using public fountains. According to Metro Water, there were 7,879 public fountains. From the census we know that 129,360 households were getting water from a "tap outside their premises". That works out at roughly 16 households or 80 persons per stand post (assuming all stand posts are functioning). The stand posts get water from the same mains that supply residential consumers. If we use the same statistic of 880 litres per connection, that works out at 11 litres per capita per day.

Table 4.2　Per capita water supply for different sections of the population of Madras

Source of domestic water	Average per capita supply (litres per day)
Households with tap in the dwelling	152
Tubewells in the dwelling	125
India Mark II pumps	125
Households with a tap in the dwelling (after adjusting for unaccounted water)	94
Static tanks	38
Shallow wells within the dwelling	16
Shallow wells outside the dwelling	12
Households who use public fountain	11

Source: Information collected and processed by the author.

One limitation of these estimates is that we use 1991 figures for the number of households (in 1996). If the population growth is taken into account, all the per capita figures would be even lower. However, that does not affect the main argument here, which concerns inequality in water distribution. These figures indicate one aspect of inequity, i.e., differences in the quantity of consumption. Another aspect of equity relates to who is paying how much.

What People are Currently Paying for Water

In 1994-95, Metro Water's total expenditure was about Rs. 840 million (or about £15.5 million in 1995 prices). However, this includes expenditure on water supply as well as sewerage. In the absence of break-up, if we assume that half of this amount is allocated to water supply, the average cost per thousand litres works out at Rs. 2.66 (approximately US$0.07 at the

October 1996 exchange rate). In comparison, the tariffs charged for metered consumers and the effective average cost for others are shown in Table 4.3.

Table 4.3   Average cost of water faced by various users in Madras

User category	Cost (Rs.)[a]	Quantity	Units	Average cost per 1,000 litres (Rs.)
A. Metro water: metered				
Industrial				25.00
Commercial				10.00
Public authorities				10.00
Bulk Consumers				20.00
B. Metro Water: Un-metered				
Industrial	1,771,670	47	MLD[b]	0.10
Commercial	24,255, 755	9	MLD	7.38
Public authorities	701,275	24	MLD	0.08
Domestic Non-residential	5,775,520	6	MLD	2.64
Domestic residential	64,660,930	97	MLD	1.83
C. Static tank users	0.25	20	litres	12.50
Vendors (from static tanks)	1.00	20	litres	50.00
D. Private Sector				
Private tankers	350	12,000	litres	29.17
Mineral water	15	12	litres	1,250.00

a. The exchange rate in October 1996 was Rs. 35.7 to US$1
b. Million litres per day

Sources: Cost details : (A) For metered users of Metro Water, tariffs from Metro Water regulations; (B) For un-metered users of Metro Water, cost details from Metro Water annual report (schedule H). Quantity details from

Water Balance Sheet in Figure 1. For C and D, data collected from field work and interviews.

Some poignant conclusions emerge. The low income households without any entitlements to water other than from the self regulated sector of static tanks may face an average cost of water that is nearly seven times that faced by an average household having a yard tap connection (or more than four times the average cost of supply by Metro Water). With an average cost of supply of Metro Water of Rs. 2.66 per 1,000 litres, unmetered industrial users are paying Rs. 0.10 per 1,000 litres, or less than 4% of cost. This is a somewhat disconcerting finding.

Water Entitlements

The foregoing discussion indicates the various dimensions of the water supply problem in Madras needing to be addressed by policy. Property rights issues can be brought into the discussion using Sen's entitlements concept. The amount of water that a person actually consumes depends on the entitlement set E_i of person i. This E_i "can be characterised as depending on two parameters viz., the endowment of the person and the exchange entitlement mapping" (Sen, 1981, p. 45). Given that water is a bulky commodity, and given the limitations of trade in water, the endowment portion rather than exchange entitlements determines the amount of water that a person gets in a system. Water endowment can be defined in terms of average per capita water provided by each source. If person X has well as a source, we know such person's endowment is 16 litres per capita per day (lpcd). If person Y has a yard tap from Metro Water, the endowment is 94 litres per capita per day, and so on. These issues have been discussed in detail elsewhere (Anand and Perman, 1998). It may suffice here to conclude with a presentation of the distribution of households according to entitlement levels.

Table 4.4 Water endowment levels of households in Madras (%)[a]

Water endowment (lcpd)[b]	Mean monthly income in Rs.					
	2,000	3,000	4,500	8,000	20,000	Total
0	40.0	34.1	9.4			17.6
16	35.0	31.8	18.8	36.4	15.8	28.4
94	10.0	13.6	18.8	6.1		10.8
110	5.0	15.9	18.8	9.1	31.6	15.5
125	10.0	2.3	6.3	6.1		4.7
219		2.3	28.1	33.3	52.6	20.9
235				9.1		2.0
Total	100.0	100.0	100.0	100.0	100.0	100.0

a. Shaded cells refer to median class (of endowment) in each column.
b. Litres per capita per day.

Source: Madras Household Survey, 1996.

In general, and as may be seen from median endowment levels, as income increases, water endowment increases. A linear regression analysis yielded the following result (t- values shown in parenthesis):

$$\text{WATENDOW} = 43.438604 + 0.004437 * \text{MEANINC}$$
$$\qquad\quad (4.986) \qquad\quad (4.357)$$

   R square: 0.115

Completion of a project to increase the total quantum of water supplied to the city *per se* does not affect the first two rows of Table 4.4 because those households are not connected to the water supply system and do not draw the benefits. Given that 75% of the lowest income group and 65% of the next income group do not have a connection, any water project that does not provide them with a connection does not affect their endowment. Hence, assertions that "the poor would particularly benefit" cannot be sustained.

Conclusions

A basic service such as water supply, something taken for granted in the western world, can be a complex issue in any city in the developing world. People in such cities seem to face two different markets and almost by chance get allocated to one of them. If they are among the lucky ones to have a connection, they face a regulated monopoly of the state, where prices are politically determined without relation to costs. The others face a market that varies from an oligopoly to competition depending on how active the regulators are and how powerful the law is.

The chapter presents some information on the nature of inequality in distribution of water and the relative prices faced by different categories of households. A policy that focuses mainly on increasing water availability per capita without altering the entitlements would not improve the well-being of the lowest income groups, but in fact increase the inequities in the distribution. Entitlements need not necessarily mean only private ownership of the water tap connection. There can be community based institutions with collective ownership.

The monopoly nature of water boards needs to be changed. Well-functioning markets are needed, but they are not a panacea to all water supply problems. Public sector institutions can play an important role of not only being a regulator, but also a market leader. But, when it comes to the lowest income groups, public sector water boards can and have to play an important role, for instance in developing appropriate property rights and protecting them (Anand, 1997). For this, the public sector water boards need to be strengthened in some respects, while their role needs to be restricted or re-oriented in others. Reforms would be needed, including institutional strengthening, including "corporatisation" of the organisation, decentralisation of functions and meeting the consumers from time to time; creating the needed capacity to enable water boards to function as service industry rather than engineering departments; making it necessary for the organisation to make a public report (at least) once a year or more frequently on both technical and financial indicators; and setting up an independent body to monitor the quality of water (chemical and bacteriological) as supplied in different parts and to publish periodic reports in news papers, etc.

In Madras, voluntarily, Metro Water brought out a "Citizen's charter" in March 1998, which is a welcome move. It is necessary to continue in this direction. The need for understanding the functioning of water markets and the need to avoid "excessive" regulation which could be counter-productive is also clear.

Note

1   See, for instance, MMDA's Master Plan, where a comparison is made with Bangalore (90 litres per capita per day), Bombay (150 litres per capita per day), Calcutta (190 litres per capita per day), Delhi (160 litres per capita per day) and Pune (275 litres per capita per day) (MMDA, 1995, p. 95).

References

Anand, P. B. (1997), "Urban Environment and the Public Sector : Inequity and Institutions in the Developing World", paper presented at the ESRC Development Economics Study Group conference on "The Role of the Public Sector", University of Reading, Reading.

Anand, P.B. and Perman, R., "Preferences, Inequity and Entitlements: Some Issues from a CVM Study of Water Supply in Madras, India", forthcoming in *Journal of International Development*.

Central Ground Water Board (1993), *Ground Water Resources and Development Prospects in Madras District, Tamil Nadu*, Central Ground Water Board, Southern Region, Hyderabad.

Chidambaram G., (1991), *Database of Madras Metropolitan Area*, volume V, Madras 2011, MMDA-TRF Research Programme, MMDA, Madras.

Dattatri G. and Anand, P. B. (1991), *Madras 2011: Policy Imperatives. An Agenda for Action*, Theme Paper, Madras 2011, MMDA-TRF Research Programme, MMDA, Madras.

Institute for Water Studies (1994), *Environmental Assessment of Madras Basin*, Interim Report, Government of Tamil Nadu, Madras.

Madras Metropolitan Development Authority-MMDA (1995), *Master Plan for Madras Metropolitan Area, 2011* (draft), MMDA, Madras.

Madras Metropolitan Water Supply and Sewerage Board-Metro Water (1995), *Annual Report 1994-95*, Metro Water, Madras.

Sen, A. K. (1981), *Poverty and Famines: An Essay on Entitlement and Deprivation*, Clarendon Press, Oxford.

Srinivasan, S. (1991), *Water Supply and Sanitation in Madras Metropolitan Area*, volume 3, Madras 2011, MMDA-TRF Research Programme, MMDA, Madras.

5 Environmental Management in the Cape Metropolitan Area

AMANDA YOUNGE

Introduction

This chapter gives an overview of current environmental management issues in South Africa, with particular reference to the Cape Metropolitan Area. It notes that, while there is a growing awareness of the importance of good urban environmental management, the emphasis remains at the policy level and falls short on implementation.

As a result of apartheid policies and rapid urbanisation, South Africa's cities exhibit severe disparities in environmental quality, are highly inefficient and do not serve the needs of the majority of the population. The country's population, while being only 55% urbanised, is growing at 2% per annum. The rate of urbanisation has been 5% per annum.[1] Because the policy of the previous government was to limit urbanisation, planning to address urbanisation constructively is in its infancy and is ill-equipped to address the existing demands.

Urban development patterns have segregated the population on the basis of race and income, and many of the costs of urban inefficiency are borne by the lower-income groups. The poor are typically located furthest from employment, and transport costs are high. Racial local government structures have had inequitable access to revenue sources, resulting in massive service disparities between areas, with low-income neighbourhoods being inadequately serviced as well as lacking in basic services such as housing, water, sanitation and access to electricity. Pollution, chemical contamination and illegal dumping are rife. Formal housing for the poor is often located nearest to environmental hazards such as landfill areas and polluting industries. Uncontrolled informal settlement in floodplains has led to flooding problems which are prohibitively expensive to resolve.

As yet, the issue of environmental sustainability and its relationship to good urban management is poorly understood. Environmental issues are still very much seen as wildlife issues.

51

In cities, popular campaigns against apartheid particularly during the 1980s focused on demands to improve urban living conditions. Linked to this was the demand for the removal of racial local government structures, especially the Black Local Authorities.[2] Civic organisations mobilised themselves against voting for these councils and, through rent and service charges boycotts, undermined their financial base. Demands for housing and services were coupled with demands for a participatory, democratic and non-racial system of local government to replace the segregated system. With the demise of the apartheid state, the newly-established non-racial democratic government was committed to addressing the urban crisis.

In October 1997, the Government of National Unity produced an Urban Development Framework for the country.[3] It views urban settlements as becoming, by 2020, spatially and socio-economically integrated centres of economic and social opportunity and of vibrant urban governance managed by democratic, efficient, sustainable and accountable local government in close cooperation with civil society. Cities are expected to become environmentally sustainable.

The document views improved environmental management as an integral part of the planning and development process. It argues that housing, planning, and services have a direct bearing on environmental quality and the health and well-being of urban residents, and it encourages all municipalities to embark on Local Agenda 21 programs, and to integrate these with local planning and development initiatives. Sustainability and environmental quality indices are being developed for the national monitoring of cities.

Despite all the progressive provisions on environmental protection of the 1996 Constitution, South Africa still lacks a coherent national and regional policy on the environment, and effective environmental management is obstructed by fragmented jurisdictional, legislative, regulatory and fiscal responsibilities.[4] If legislation is reasonably adequate, implementation and enforcement are deficient; increasing pressure for stricter protection measures is being experienced as a result of the country's re-entry into the world market.

The Environment Conservation Act of 1989 provides the Minister of Environmental Affairs and Tourism with potentially far-reaching powers. A General Environmental Policy was published in 1994 and a White Paper on an Environmental Management Policy was published in July 1997. Other national policies include policy on Conservation of Biological

Diversity and the Integrated Pollution Control and Waste Management Policy, while a draft White Paper has been recently produced and is being circulated for comment.

The 1989 Act was recently amended, giving power to local, provincial and national government to halt any development where a negative environmental impact is anticipated. The relevant authority can then force the developer to rehabilitate the site, failing which the authority can rehabilitate it at the developer's cost.[5] Regulations were recently promulgated in terms of this Act, to be implemented by the Provincial Administrations, compelling developers to assess environmental impact before embarking on new projects. Unfortunately, given their limited resources, the Provincial Administrations may have difficulty fulfilling their responsibilities.

In addition, on 2 November 1995 South Africa ratified the Convention on Biodiversity which was opened for signature at the 1992 UNCED in Rio de Janeiro. The government is bound to subscribe to the principles and articles laid down in the Convention and to implement it at all levels.

The Process of Municipal Reorganisation

A crucial aspect of constitutional change in South Africa has been the reform of local government. Apartheid local government was racially, fiscally and institutionally segregated. Urban unrest in the 1980s targeted Black Local Authorities, which were the corrupt, inefficient and unrepresentative structures established to govern Black residential areas. Reform has been directed at desegregating local government and establishing fiscally viable municipal structures.

In negotiations at national level in the early 1990s, it was agreed that local reform would begin with the establishment of national legislation to guide the process. Through the struggle against apartheid, civil society developed extensive skills and networks, having played a crucial role in transformation. To enable participation by local CBOs and NGOs, many decisions regarding local transition were devolved to local negotiating fora.

The next step was to re-demarcate municipal boundaries to establish non-racial municipalities, by integrating each functional urban area under one municipal jurisdiction; in metropolitan areas, a two-tier system of local government was established. Local elections followed on the basis of universal suffrage and a process of administrative and fiscal amalgamation has taken place. The local elections, while resulting in democratically elected councils, have had a negative impact on CBOs and NGOs, since many of their leadership have been elected as councillors.

The reform process to date has been aimed at simply eliminating racial elements from local government.   It has not attempted to transform municipal governance. This was scheduled for a subsequent phase, once the unification process is completed. The new Constitution created the space for this second phase to begin and the framework within which local government would operate in future. The next steps in the transformation process have been spelt out in the Green Paper on Local Government.[6]

The National Policy on Local Government

A discussion document entitled "Towards a White Paper on Local Government"[7] was produced in May 1997 to set out the basis for a new approach to local government. It argued that the promotion of integrated and coordinated development is a priority, as is the creation of a safe and healthy environment. It viewed environmental sustainability as being closed related to growth and redistribution. Arguing that the improvement of living conditions - particularly of the poorest sections of communities - cannot take place at the expense of the environment, the document concluded that environmental sustainability largely depends on the quality of urban management.

The document also highlighted the role of local government in overall governance. During the apartheid period, as local government structures became discredited, informal governance structures emerged in civil society, taking over some of the responsibilities of formal local government. With the establishment of non-racial democratic local government structures after 1994, the role of structures in civil society needed to change.

With almost no tradition of legitimate local government, many communities found it difficult to understand what role could be played by a municipal council and how they could participate in its decision-making. The establishment of participatory local government assumes that organs of civil society understand how to interact with it, but in fact many councillors drawn from those same communities have had difficulty understanding how to be effective councillors, as opposed to civic activists. Their problems have been compounded by the inertia and inflexibility of local bureaucracies, unused to a participatory style of government. In this context, the 1997 Green Paper on Local Government set out a vision for developmental local government which should plan and manage development in an integrated and sustainable manner.[8]

Direct support for improved environmental management can also be found in Chapter 1 of the Development Facilitation Act of 1995, which set out general principles for all land development.[9] These include the requirement that policy, administrative practice and laws should promote efficient and integrated (urban and rural) land development, aiming at the integration of the social, economic, institutional and physical aspects of land development. Policies and laws are expected to promote a diverse combination of land uses; discourage the phenomenon of "urban sprawl" and contribute to the development of compact cities; contribute to the correction of the historically distorted spatial patterns of settlement and to the optimum use of existing infrastructure in excess of current needs; and encourage environmentally sustainable land development practices and processes. They should also promote sustainable land development within the country's fiscal, institutional and administrative means and ensure the safe utilisation of land by taking into consideration factors such as geological formations and hazardous undermined areas.

The Local Government Transition Act (Second Amendment Act) of 1996 required every municipality (including metropolitan councils) to prepare and annually revise an integrated development plan observing general principles contained in Chapter 1 of the Development Facilitation Act.[10] Human resource planning and capital development programs must be aligned with the integrated development plan. Crucial to the integrated development plan are the requirements that a council must give priority to the basic needs of its community, regularly monitor and assess its performance against its integrated development plan, and annually report to, and receive comments from, its community. This reflects directly community demands for a more participatory model of local government.

Environmental Planning and Management in the Major Metropolitan Areas

There are three major metropolitan areas in South Africa, namely, Greater Johannesburg/Gauteng Province, Greater Durban, and the Cape

Metropolitan Area. They have a history of fragmented jurisdictions, massive urbanisation, racial segregation and vast disparities in wealth and service levels. A metropolitan level of local government has been established for the first time, and with this has come the opportunity for environment, health and development issues to be addressed on a more coordinated and integrated basis. Recently, all three areas have become involved in Local Agenda 21 and Healthy Cities programs and have established a "Three Cities Network" to support one another in this process. Increasing attention is being given to environmental issues, with the priority on "brown" environmental issues - primarily addressing basic service needs.

Metropolitan Cape Town

The Cape Metropolitan Area (CMA) has a unique natural environment, famous for Table Mountain (the Peninsula Mountain Chain) and the coastal landscape. The South Western Cape is home to the Cape Floral Kingdom, comprising some 8,500 species of which 70% occur nowhere else on earth.[11] The region's climate is Mediterranean, subject to severe South-easterly gales during the summer and North-easterly storms in winter. In winter, temperature inversions are common over the Cape Flats, the sandy low-lying area linking the Peninsula to the mainland, causing ambient air quality standards to be very poor as the pollution from motor vehicles and local industries remains trapped in a layer of surface air which is cooler than the layers above it.

The city itself originated as a colonial port city, and it has a particularly eccentric structure due to both the surrounding topography and the impact of apartheid policies. Former Black Local Authorities, denied access to property rates income from wealthy parts of the city and dogged by legitimacy problems, were unable to provide adequate services to their residents.

While high-income urban sprawl consumes land containing valuable resources, rapid urbanisation has increased the tendency for the poor to be concentrated in dense informal settlements on the urban periphery. This has resulted in massive urban inefficiency and high commuting costs. The area generally occupied by the poor is known as the Cape Flats, which has sandy soil, a high water table, poor drainage, vulnerability to flooding and a fragile, degraded ecosystem. Climatic conditions conspire to make it the part of the city least suitable for human habitation. The burning of wood for fuel worsens pollution problems resulting from vehicle and industrial emissions.

Health problems for the city's poor have been compounded by the collapse of service provision in the townships in the late 1980s precipitated by rent and service charges boycotts. Already inadequate municipal services disintegrated, resulting in problems such as extensive blockages in stormwater and sewerage systems, flooding, and an almost total absence of refuse removal. As a consequence, the city has one of the world's highest tuberculosis rates. Poor catchment management, polluting urban and agricultural runoff and inadequate solid and liquid waste disposal (particularly in low-income areas) has resulted in poor water quality in rivers, water bodies and coastal waters.[12]

Municipal Reorganisation in the Cape Metropolitan Area

The prospect of municipal desegregation in metropolitan Cape Town struck fear into the hearts of its affluent citizens: that sharing municipal revenues between rich and poor areas and taking responsibility for the welfare of all citizens would bankrupt the "haves". Municipal unification at a metropolitan level was opposed by the governing party in the Western Cape Provincial Legislature, the National Party.

Unable to resist the establishment of a metropolitan council, the National Party shifted its attention to ensuring that the metropolitan boundary would exclude two satellite towns which are functionally part of the metropolitan area. This has meant that the effective management of the whole functional metropolitan area cannot be undertaken by the metropolitan council, whose jurisdiction is too limited. It can now only be achieved through joint agreements with neighbouring district councils, or by the Provincial Government, a less than satisfactory situation in the light of the area's high levels of urbanisation.

Six new metropolitan local councils were established where there had been 69 local government structures of various kinds in the past. Unlike the physical problems encountered in demarcating Johannesburg and Durban, it was possible to meet the requirements of the legislation to unite Black and White local authorities in this process, despite opposition from conservative sectors. Such opposition led to an unsuccessful Electoral Court challenge to the recommendations of the Demarcation Board, delaying local elections in the area.

When the challenge failed, conservative resistance to political amalgamation shifted to resisting administrative amalgamation. For instance, the National Party-controlled municipality of Tygerberg, one of the six metropolitan local councils, has preserved administrative segregation within its area of jurisdiction, setting up separate area-based administrations whose boundaries broadly follow those of the former segregated authorities. If the financing of the different areas remains discriminatory, as in the past, limited progress will be achieved in terms of equity in service provision in this municipal area, which includes Khayelitsha, a large low-income dormitory settlement, which houses approximately 300,000 people.

A Strategic Management Plan was introduced by local government in the CMA in 1994 to address some of the problems arising from the collapse of services in Black areas. It was conducted on a collaborative basis, prior to the amalgamation of local authorities. White Local Authorities, funded by local government levies, established a special program to restore services to levels prior to their collapse. This was an initial step towards rectifying the situation. The program ran over three years and was geared at rebuilding management and administrative capacity, unblocking stormwater drains and sewers, reconstructing roads and repairing damaged water supply systems. It is now coming to an end and service provision in low-income areas is becoming increasingly integrated into the mainstream line functions of the newly-established councils. The program was successful in achieving its objectives, and has improved environmental health and water catchment management, but it has by no means met the servicing needs of residents in these areas. The next step is to extend services to unserviced informal housing areas, and to improve service levels overall.

In addressing this challenge, the newly-elected councils must deal with the persistence of the "culture of non-payment". Although this originated in politically-motivated rent and service charges boycotts, non-payment has now become an issue in many parts of the city, poor and affluent, residential and commercial. Former Black municipal administrations had inefficient or non-existent billing systems, were riddled with corruption, provided no services to speak of and were not able to enforce payment. During the early 1990s, former White municipal administrations began to find that increasing numbers of people in their areas were following suit by defaulting on water, electricity and rates accounts.

The new municipalities have until recently lacked the political will to enforce cost-recovery strategies. However, their ability to provide services in the future largely depends on how effective their revenue collection capabilities are. Councils are attempting to improve cost-recovery through programs such as "Interim Community Charges" instituted recently by the Cape Town Municipality, one of the six metropolitan local councils.

Environmental Management in the CMA

Because of the Cape's scenic beauty and the importance of the environment to tourism and local economic development, as well as the disparities between rich and poor areas, there is a high level of awareness of the need for better environmental management in the metropolitan area. Political priorities focus on upgrading low-income areas, but the links between this process and overall issues of environmental sustainability are not emphasised.

Initiatives to create a framework for metropolitan growth management date from 1995, when the Cape Metropolitan Council-CMC approved a document entitled "Principles for Planning and Development in the Cape Metropolitan Area".[13] This document emerged from a process of wide consultation with key metropolitan stakeholders, containing principles of equality of opportunity, social justice and sustainable development as a basis to guide for future planning and development.

The CMC then prepared a Metropolitan Spatial Development Framework in 1996.[14] It argued that containment of urban growth was critical to ensuring both an efficient city and effective environmental conservation in the region. To achieve this, provision had to be made for increasing the density and intensity of activities in the city, especially through fostering the development of high-intensity "activity corridors" linking under-serviced low-income settlements on the urban periphery with better serviced areas, bringing higher order activities to dormitory townships and improving the public transport system on which these corridors would be focused.

Despite its merits, the framework is being challenged by trends in the urban land market. Problems have already emerged as the locational decisions of larger investors have preferred not to follow the planners' directives. In a city with limited economic growth, little resistance is posed to the locational preferences of major developers, however inappropriate these may be from the viewpoint of the urban structure.

A second implementation problem relates to the lack of adequate tools to manage the urban edge. Already the demarcation of the metropolitan area to exclude satellite towns has weakened the CMC's ability to manage the urban edge effectively. Urban growth in these towns impacts on a highly sensitive natural resource base and will drastically increase the costs of urban management.

The new CMC is charged with preparing and ensuring compliance with an integrated metropolitan environmental management policy and strategy. An Environmental Planning Department has been established in the planning directorate, which will be responsible for doing this work and coordinating the Local Agenda 21 and Healthy Cities initiatives in the metropolitan area. Critical areas of focus will include the management of the new Cape Peninsula Protected Natural Environment (the Peninsula Mountain chain), the management of the urban edge and the holding of an environmental EXPO in the metropolitan area.

The role of the Coalition for Sustainable Cities should be mentioned. Created in 1993 as a working group of the Cape Town-based environmental and service network, the Green Coalition, which is itself a regional arm of the national Environmental Justice Networking Forum, the CSC defined its mission as being to contribute to the establishment of urban policy that forms a basis for environmental sustainability. To support this, the CSC has proposed that an environmental advisory forum be established by the newly established metropolitan council to advise it on environmental management policy and to provide resources to activities that facilitate broader partnership in the implementation and monitoring of these policies.[15]

Conclusion

South African cities are faced with critical environmental management problems and significant resource constraints. The preconditions to address these problems have been established in the Constitution, in government policy and in legislation. However, the Constitution has been likened to a Rolls Royce without petrol. The space that has been created needs to be used by civil society and government at all levels to improve environmental management. Indications are that the NGO sector has the capacity to support civil society in this process.

However, the prioritisation of environmental issues remains a political choice, and unless there is a clear understanding of the links between improving people's day-to-day living conditions and the need for effective environmental management, it may be difficult to ensure a sustainable future.

Notes

1   Ministry of Housing (October 1997), 'Living Cities: Urban Development Framework', Government of the Republic of South Africa, Executive Summary.
2   Black and White when capitalised refer to the racial terms used by the apartheid government.
3   See note 1.
4   Constitution of the Republic of South Africa 1996 (Act 108 of 1996), Government of the Republic of South Africa; Section 24 of the Bill of Rights.
5   Environment Conservation Act of 1989 (Act 73 of 1989), Government of the Republic of South Africa, section 31(a).
6   Ministry for Provincial Affairs and Constitutional Development (October 1997), Green Paper on Local Government, Government of the Republic of South Africa.
7   Ministry for Provincial Affairs and Constitutional development (1997), 'South Africa's Local Government: A discussion document: Towards a White Paper on Local Government in South Africa', Government of the Republic of South Africa.
8   See note 6.
9   Development Facilitation Act (Act 67 of 1995), Government of the Republic of South Africa.
10 Local Government Transition Act (Second Amendment Act)(Act 97 of 1996), Government of the Republic of South Africa.
11 See Quick (1997) and Jones (1997) for more information.
12 Environmental Unit, Town Planning Branch, 'Environmental Evaluation for the Cape Metropolitan Area: City of Cape Town', August 1993.
13 Cape Metropolitan Council, (1995), 'Principles for Planning and Development in the Cape Metropolitan Area'.
14 Regional Planning Branch (1996), 'Metropolitan Spatial Development Framework: Cape Metropolitan Council'.
15 Coalition for Sustainable Cities (1997), 'Policy options towards a sustainable Cape Town', working document.

References

Quick, A.J.R. (1997), 'Cape Town: State of the Environment, City of Cape Town', discussion document, Cape Town.

Jones, L. (1997), 'Comments on Theme 4 of the Cape Town Olympic Bid File', discussion document, Cape Town.

6 The National Urban Parks Programme in Santiago, Chile

SERGIO F. LEÓN BALZA

Introduction

This chapter discusses the National Urban Parks Programme which has been undertaken by Chile's Ministry of Housing and Urban Development since 1992. Following a general presentation of the programme, it discusses its main results and shortcomings in stimulating local initiatives and long lasting community commitment. Finally, the chapter discusses some strategic issues and options for future action.[1]

In 1992, an Urban Parks Programme was launched to meet the recreational needs of the urban poor particularly in the Santiago Metropolitan Area, where a series of sites of significant size have been reserved as parks since the city's 1960 land use plan. Greater Santiago accounts for 40% of the country's population and is well-known for its high rates of air pollution. Given the local government's incipient resources, most of these parks were largely abandoned, when they were not used as informal rubbish dumps and gathering places for drug addicts. Most of Santiago's green space is concentrated in the business areas and in the high-income boroughs or municipalities. Only a very small percentage of the population has had access to quality green space within their own boroughs.

From the outset the programme received direct support from the Minister then in office, which has certainly opened doors within the Ministry's bureaucratic structure and accelerated many administrative procedures. The programme has also attracted international financial support, which has been used to fund annual technical conferences, contract foreign consultants, and facilitate studies and personnel training. However, its shortcomings soon appeared and it is currently being re-assessed; strategic problems have been recognised, calling for urgent co-ordination with local authorities, the private sector and the community.

The New Democracy and the Green Space Deficit in 1992

The 1990 transition to democracy in Chile has been followed by increasing social demands. The military regime that ruled the country from 1973 to 1989 designed a development policy centred in the market, often categorised as "savage capitalism" for its excessive confidence in the possibilities of economic growth. The social policy of those years was based on the deconcentration of local authorities' resource funding and on the creation of subsidies in areas such as health, housing, education and nutrition. As previously discussed, these subsidies were used as mechanisms of manipulation and political control, rather than as a means of real social improvement.[2]

From the viewpoint of urban development, the years of military rule brought about very small-sized social housing units and evident spatial segregation of the different social strata. Such segregation was worse for low-income households, because the official housing policy aimed only to provide houses without infrastructure and social services. It was worsened further by the unstable insertion of that particular fraction of the labour force into the market economy.[3] The 1979 urban policy abolished the standing city limits, provoking a series of negative impacts on the national urban structure.

The official policy of the 1990s emphasises serving the poorer strata of the population, thus reviving a Chilean tradition of directing a significant proportion of the GDP to social programmes.[4] As an immediate result of the changing national policies, housing and urban development public programmes have diversified and multiplied.[5]

Green Spaces in Santiago: The 1992-1994 Situation

It is in this context that the political decision to initiate a National Programme of Urban Parks directed to the low-income sectors has to be understood. The programme's general objectives are to provide a more equitable distribution of urban services and to respond to other environmental concerns such as reducing Santiago's high air pollution levels and improving the quality of the urban structure. Such a political decision was made at a moment when macro-economic variables did not yet show today's stable increase in real incomes, allowing for the forecast of a rising demand for amenities and green spaces in urban areas. An idea of the distribution of green areas in Santiago is given in Table 6.1.

Table 6.1   Publicly-accessible green areas in Greater Santiago

Municipality	Number of areas	Surface (m2)	Share of total surface (%)
Renca	40	2,171,048	11.00
Santiago	57	1,964,028	9.95
Las Condes	230	1,734,753	8.79
Recoleta	53	1,436,896	7.28
Ñuñoa	90	1,166,583	5.91
Vitacura	94	845,025	4.28
La Granja	99	805,938	4.08
La Florida	282	709,309	3.59
La Reina	51	660,114	3.34
Conchalí	90	624,402	3.16
San Bernardo	78	579,916	2.94
Providencia	77	564,536	2.86
Quinta Normal	48	559,974	2.84
Estación Central	80	553,071	2.80
Maipú	181	531,641	2.69
La Cisterna	24	440,730	2.23
Lo Prado	74	437,190	2.21
Macul	55	429,810	2.18
Cerro Navia	71	344,729	1.75
San Joaquín	63	342,142	1.73
Pudahuel	35	320,595	1.62
Lo Barnechea	51	314,422	1.59
La Pintana	98	300,374	1.52
Puente Alto	94	298,106	1.51
P.A. Cerda	45	248,920	1.26
San Ramón	56	231,462	1.17
El Bosque	27	229,912	1.16
San Miguel	25	209,255	1.06
Lo Espejo	23	171,256	0.87
Independencia	37	165,395	0.84
Quilicura	40	112,599	0.57
Peñalolén	33	109,550	0.55
Cerrillos	22	77,903	0.39
Huechuraba	22	47,998	0.24
Subtotal	**2,445**	**19,739,582**	**100.00**
Parque Metropolitano	1	7,120,000	
Total	**2,446**	**26,859,582**	

Source: CEC Consultores (1992) and information provided by Ministerio de Vivienda y Urbanismo (MINVU) and the 'Programa de Percepción Remota y Sistemas de Información Geográfica', Pontificia Universidad Católica (PPR).

The most centrally-located municipality (i.e. borough) of Santiago, where the central business district and office areas of the city are located, and the middle- and high-income residential areas of Las Condes, Ñuñoa and Vitacura, jointly contain 28.93% of the total accessible green space (cf. Table 6.1). Regarding the distribution of green spaces which actually receive regular maintenance, the situation is even more dramatic. Table 6.2 shows that, excluding the Metropolitan Park (which by itself amounts to almost 38% of the total available green space in the city), 30.81% of the total area of maintained green spaces is concentrated in four of the 34 municipalities that comprise the city: Las Condes (6.2%), Santiago (13.21%), Providencia (3.28%) and Ñuñoa (8.12%), which are all areas of predominantly middle- to high-income households and the central business district of the city. When combined, the same boroughs account for 62.92 % of total expenditure on green space maintenance, leaving the remaining 30 boroughs, none of which counts with more than 3% of total green space in the city, with the remaining 37.08% of the global budget. One can only infer that the low budget allocated to green space maintenance in the poorer districts is also accompanied by the lowest quality of maintenance.

Table 6.2   Maintenance of publicly accessible green areas in Greater Santiago

Municipality	Area maintained (m2)	(%)	Share of annual budget (US$)	(%)	Expenditure (US$/m2)
Las Condes	720,456	6.20	2,778,088	28.62	0.321
Santiago	1,535,618	13.21	1,852,445	19.08	0.101
Various (MINVU)*	246,000	2.12	508,440	5.24	0.202
Providencia	381,621	3.28	900,294	9.27	0.197
Ñuñoa	943,778	8.12	577,455	5.95	0.051
Conchalí	339,729	2.92	539,956	5.56	0.132
Maipú	116,177	1.00	300,493	3.10	0.216
Macul	195,253	1.68	262,148	2.70	0.112
Est,Central	253,976	2.19	230,341	2.37	0.076
San Miguel	95,070	0.82	217,324	2.24	0.190
La Florida	267,371	2.30	198,109	2.04	0.062
La Pintana	170,518	1.47	158,673	1.63	0.078
La Reina	304,650	2.62	155,585	1.60	0.043
San Joaquín	212,298	1.83	147,450	1.52	0.058
La Cisterna	238,917	2.06	144,338	1.49	0.050

Renca	1,784,084	15.35	135,539	1.40	0.006
Peñalolén	49,768	0.43	108,218	1.11	0.181
Pudahuel	182,026	1.57	101,578	1.05	0.046
Cerro Navia	155,880	1.34	101,217	1.04	0.054
San Ramón	52,132	0.45	83,069	0.86	0.133
La Granja	565,856	4.87	80,629	0.83	0.012
Lo Espejo	83,036	0.71	69,638	0.72	0.070
Quilicura	77,677	0.67	53,041	0.55	0.057
Q.Normal	306,423	2.64	3,523	0.04	0.001
Puente Alto	135,430	1.17	-	0.00	-
Lo Barnechea	142,846	1.23	-	0.00	-
Independencia	39,522	0.34	-	0.00	-
San Bernardo	347,513	2.99	-	0.00	-
Cerrillos	32,685	0.28	-	0.00	-
Huechuraba	21,805	0.19	-	0.00	-
El Bosque	138,568	1.19	-	0.00	-
Lo Prado	167,256	1.44	-	0.00	-
Recoleta	611,223	5.26	-	0.00	-
Vitacura	591,689	5.09	-	0.00	-
P.A. Cerda	114,754	0.99	-	0.00	-
Total	**11,621,605**	**100.0**	**9,707,591**	**100.0**	0.107

* Figures for the Ministry of Housing and Urban Development (MINVU) do not include the Metropolitan Park (where about half the surface is accessible).
Sources: 'Catastro de Areas Verdes del Area Intercomunal de Santiago', CEC - PPR/UC, 1992; calculations based on Central Bank of Chile's consumer price index.

Regarding the figures of maintained green space on a per person basis, the differential access by income groups to green spaces becomes even clearer. Table 6.3 shows that, with the exception of those boroughs where the Ministry has created new green areas or where there are hills considered as "official green areas" (which may not be maintained), quality and accessible green spaces are a prerogative of the wealthy.

Table 6.3   Maintenance of public green areas in Greater Santiago, 1992

Municipality	Population	Area under maintenance (Ha)	Distribution of maintained areas (%)	Area maintained per person (m2)
Renca	129,173	178.41	15.68	13.81
Santiago	202,010	153.56	13.50	7.60
Vitacura	78,010	59.17	5.20	7.58
Ñuñoa	165,536	94.38	8.30	5.70
La Granja	126,038	56.59	4.97	4.49
Recoleta	162,964	61.12	5.37	3.75
Las Condes	197,417	72.05	6.33	3.65
La Reina	88,132	30.47	2.68	3.46
Providencia	110,954	38.16	3.35	3.44
Lo Barnechea	48,615	14.28	1.26	2.94
Q.Normal	115,964	30.64	2.69	2.64
La Cisterna	94,732	23.89	2.10	2.52
Conchali	153,089	33.97	2.99	2.22
Quilicura	40,659	7.77	0.68	1.91
San Joaquin	112,353	21.23	1.87	1.89
San Bernardo	188,580	34.75	3.05	1.84
Est.Central	142,099	25.40	2.23	1.79
Macul	123,535	19.53	1.72	1.58
Lo Prado	110,883	16.73	1.47	1.51
Pudahuel	136,642	18.20	1.60	1.33
San Miguel	82,461	9.51	0.84	1.15
La Pintana	153,586	17.05	1.50	1.11
Cerro Navia	154,973	15.59	1.37	1.01
P.A. Cerda	128,342	11.48	1.01	0.89
El Bosque	172,338	13.86	1.22	0.80
La Florida	334,366	26.74	2.35	0.80
Lo Espejo	119,899	8.30	0.73	0.69
Puente Alto	254,534	13.54	1.19	0.53
San Ramón	101,119	5.21	0.46	0.52
Independencia	77,539	3.95	0.35	0.51
Cerrillos	72,137	3.27	0.29	0.45
Maipú	257,426	11.62	1.02	0.45
Huechuraba	61,341	2.18	0.19	0.36
Peñalolén	178,728	4.98	0.44	0.28
Total	**4,676,174**	**1,137.58**	**100.00**	**2.58**

Sources: See Table 6.2.

The launch of the National Urban Parks programme has already altered the balance of green areas provision in the city, providing 130 new hectares in 1996. This raised by more than 20% the total maintained green space, that is, an amount of new green land equivalent to that already present in boroughs of richer municipalities such as Las Condes, Providencia and Lo Barnechea grouped together.

Basic Concepts and Strategies of the National Urban Parks Programme

The National Urban Parks Programme sets out to promote better quality of life and environmental improvements in poor sectors of Chile's urban areas. Having placed initial emphasis in the city of Santiago, it was later oriented to urban areas in the rest of the country. The new parks initially followed European design trends, being conceived as recreational and sports facilities; they have gradually incorporated concepts related to sustainability and the protection of natural landscapes.

In terms of strategy, the programme has accentuated the necessity of involving participatory design procedures, actively incorporating local officials and the community to create an identity and a sense of shared responsibility for the investments made. A second element has been the search for an integrated approach to investments, combining the efforts by local boroughs and the community to improve the structure of local urban landscapes. This has meant support for complementary projects such as pedestrian walkways and tree planting.[6] A third strategic issue has been the promotion of what the Ministry has called an "ecological approach", by utilizing underground water for irrigation, restricting the lawn surface to reduce water consumption, using a moderate use of illumination and promoting native flora and natural designs wherever possible and practical.

Investment Strategy

Regading the selection of sites, the programme has given priority to land already reserved as green areas in Chile's 'Planes Reguladores' (Master Plans) formulated at a metropolitan and municipal or borough level. In the city of Santiago, though reserved since 1960, these sites could not be developed due to budgetary restrictions of the different local administrations. A basic locational strategy has been to search for spatial equity, distributing the Ministry's projects across the cardinal points of the periphery of the city. The programme also promotes the re-utilisation of river banks as well as landfills. In Santiago, the Ministry has so far created two parks over landfills, André Jarlán (30 has) and La Castrina (9.3 has), and is currently building a third one.

The new parks are to be maintained through a special service. As from 1997, the Ministry is requiring a greater commitment from local authorities, including the assurance of at least 50% of maintenance costs as a prerequisite to create new parks within their territory.[7]

Results Obtained During the Period from 1992 to 1996

According to the Urban Development Division of the Ministry, 103 projects were included in the programme during the period 1992-96. This figure involves a certain amount of double accounting as different phases of an intervention in a park are counted as separate projects. The figure also refers planned projects which are not yet implemented. Nonetheless, the overall results of the programme have so far been positive, with 33 parks created (representing over 200 hectares of new green areas) and over US$ 32 million in investment.

Table 6.4   Investment in parks in Greater Santiago, 1992-1997

Year	Investment (millions of US$)	Area open to the public (Has.)
1992	2.75	-
1993	4.27	26.50
1995	7.29	65.80
1996*	6.00	110.43
1997*	7.60	-
Total	**27.91**	**202.73**

* Estimates
Source: Ministry of Housing and Urban Development (MINVU).

When seen in the context of the National Urban Parks Programme, Greater Santiago has benefited particularly both in terms of areas rehabilitated or created and the volume of investment. The rest of the country has, nevertheless, experienced a gradual increased participation in the programme.

Advantages and Disadvantages of Central Government Operation

The programme's achievements may be attributed to the effectiveness of central government actions, including those of an advisory committee and the energetic involvement of a Minister who is personally committed to the programme. This provides clear evidence that central government can indeed be an effective counterpart to international support - and an efficient pivot for levering impacts at a national level - when there is political will.

International support has come from sources as diverse as the UN Urban Management Programme for Latin America and the Caribbean, the European Union and the British Government's Darwin Fund for the Survival of the Species. Such financial support has helped to promote the initiative, enhance its exposure to the media and stimulate related initiatives originating from NGOs and municipal governments. These have developed actions similar to those of the Ministry, and they have even exceeded its range of action especially at the level of neighbourhoods.

But the programme's shortcomings soon appeared. Several strategic problems have been recognised, calling for urgent and efficient co-ordination with local authorities as well as of the private sector and the community. Some of the problematic issues include the diversification of funding sources for the creation and administration of parks both in Santiago and other cities; the effectiveness and continuation of community participation in the projects; and the incorporation of natural habitats and native flora in a more extensive manner as a means to reduce construction and maintenance costs, increase biodiversity and contribute to obtaining higher levels of sustainability in the urban parks projects. The programme seems to have come to a crucial phase.

Diversification of Administration and Maintenance Funds

Little has been achieved in terms of the diversification of the sources of both administration and maintenance funds, despite the recommendations and examples from other experiences.[8] Only in a few particular situations have such expectations been partly fulfilled, as is the case of the 'Parque por la Paz' in the municipality of Peñalolén, a former concentration camp of the military regime for the management of which a non-profit private corporation was formed. This organisation obtains international funding and direct support from the municipality which covers maintenance costs. Another case is that of Parque André Jarlán, in the central-southern municipality of Pedro Aguirre Cerda, where the Ministry funds almost the totality of maintenance costs, while the municipality contributes nominal payments to elderly volunteers who help maintain and protect the park from potential abuse. This programme is overtly a form of social support, rather than an alternative means of attracting fresh resources and knowledge into park management.

Limits to Community Participation

From the outset the issue of participation has been important. There have been serious attempts at generating continuous community involvement in all phases of project development, but these have been restricted due to institutional and legal obstacles.

The Ministry cannot legally engage in directly promoting social organisation, unless through indirect formulae such as especial agreements with agencies that self-finance their activities related to ministerial programmes, or through particular requirements within the social housing application systems. Experience has indicated that the best the Ministry can do is to promote participation in the design processes. This constitutes an important, but insufficient, aspect of social promotion concerning the use of open space as a means for cultural, recreational, social or even political enhancement of local communities. Local administrations do play an important role in these matters, but the traditional clientelistic relationship between them and the Ministry has not facilitated the organisation of partnerships for integrated actions.

Moreover, the Ministry's professional structure is comprised of architects, engineers and technicians from the construction sector, who are accustomed to finishing a work - be it a road, a housing programme or a park - and then leaving it on the hands of local administrations. When social workers are employed by the Ministry, they tend to restrict their role to training and mediating in the development of social programmes, leaving community development as a matter for central government.

The Incorporation of Natural Habitats and Native Flora

The challenge of incorporating natural habitats and native flora seems easier to tackle, as it relates to areas of action where central government can be more efficient, namely, technical promotion, funding studies and generating partnerships amongst different institutions. The Ministry has been successful in obtaining a general agreement amongst professionals and institutions from the community and the private sector on the benefits of the pervasive use of native flora and the recuperation and enhancement of natural habitats within the city. There is much need for scientific experimentation and commercialisation, but a trial-and-error process is already under way in several spheres of the green space design, construction and maintenance. There has been an increase in the number of private nurseries specialised in the production of native flora.

Options for the Future

Successful green space systems in cities must be supported by as wide a variety as possible in terms of forms of ownership and administration and maintenance agencies. The Ministry is working on the elaboration of a law that permits non-profit and charity organisations to be constituted amongst regional and local authorities, with the participation of community organisations and private entrepreneurs. This initiative will certainly contribute towards enabling the formulation of several forms of co-responsibility for the management of green spaces.

In the mean time, the Ministry continues to fund and promote studies, technical meetings and exchanges amongst different actors, while 'down-to-earth' community development issues increasingly remain in the hands of local administrations, social organisations and private non-profit organisations. However, other basic needs such as housing, nutrition, education and training will certainly require much of their time and special attention for a long while ahead, reducing the attention they can give to the creation and maintenance of urban parks.

Notes

1   This chapter expresses the views of the author and does not necessarily represent the institution in which he works. It draws from previous work by the author (León Balza, 1996). An exchange rate of 405 Chilean pesos per U.S. dollar is used in calculations.
2   For an analysis of the actions of the military regime in Chile, see Bitar (1988).
3   For an analysis of living conditions in poor areas during military rule, see, among others, Barrios (1985) and Campero (1987).
4   For an analysis of historical trends in Chilean social policy, see Fernández Baeza (1989).
5   For a description of recent urban policies (1990-93), see MINVU (1994).
6   I have discussed this aspect elsewhere; see León Balza (1994).
7   See MINVU (1997) for current recommendations.
8   Several international experiences were discussed during the first National Meeting on Parks and Open Space organised in Santiago in September 1994.

References

Barrios, C. (1985), *Características del Mundo Poblacional Pobre*, ICHEH, Santiago de Chile.
Bitar, Sergio (1988), *Chile para Todos*, Editorial Planeta, Santiago de Chile.
Campero, G. (1987), *Entre la Sobrevivencia y la Acción Política*, Estudios ILET, Santiago de Chile.
CEC Consultores (1992), 'Catastro de Areas Verdes de la Intercomuna de Santiago' (mimeo).
Fernández Baeza, M. (1989), 'Las Políticas sociales en el Cono Sur: 1975 - 1985', *Cuadernos del ILPES*, No. 34, Santiago de Chile.
León Balza, S. (1994), 'Espacios Abiertos Urbanos: Conceptos, Políticas, Estrategias y Desafíos', Ministerio de Vivienda y Urbanismo, Santiago de Chile.
León Balza, S. (1996), 'El Programa Nacional de Parques Urbanos - Chile: Un caso de Mejoramiento de la Equidad Urbana', Working Paper of the Urban Management Programme for Latin America and the Caribbean (PGU-LAC) and Habitat - UNDP, Quito.
Ministerio de Vivienda y Urbanismo-MINVU (1994), 'Hacia una política urbana y de vivienda eficiente y equitativa', Appendix to the 1993 Annual Report, Santiago de Chile.
Ministerio de Vivienda y Urbanismo-MINVU (1997), 'Recomendaciones para la formulación de proyectos de parques urbanos', División de Desarrollo Urbano MINVU, Santiago de Chile.

7 The Governance of Waste Management in African Cities

J. M. LUSUGGA KIRONDE

Introduction

Africa is undergoing rapid changes. In most countries, a major population redistribution in the form of rapid urbanisation is taking place in a context of poor economic performance. Besieged by a plethora of problems, metropolitan authorities throughout the continent are unable to cope with the consequences of urbanisation. One major area where they have failed concerns waste management. In most urban areas, only a fraction of the waste generated daily is collected, and safely disposed of, by the authorities. Collection of solid waste is usually confined to a few areas, the city centre and high-income neighbourhoods, and service is usually irregular. Most urban areas never benefit from public services, having to bury or burn their waste, or to dispose of it haphazardly. Most waste is just dumped and not properly disposed of, and dumping sites are few. The sight of heaps of stinking, uncollected waste, or waste disposed of by roadsides, on open spaces, in valleys and drains is a common feature of African urban areas. Only a tiny fraction of households are connected to a sewer network or to local septic tanks, and emptying or treatment services hardly exist. Industrial waste is usually disposed of untreated into the environment.

Waste is generated by the ton in most popular areas. Law and tradition require that it be removed, and disposed of, by urban authorities. Unlike in rural areas, uncollected urban waste is a danger to health, pollutes the environment, is a nuisance, erodes civic morals and can be a major social problem. Thus, waste management is an important area of both environmental management and urban governance. This chapter looks at the problem of the governance of waste management in Dar es Salaam, the largest city in Tanzania. It is based on a study of various documents, observations carried out throughout the city, as well as on interviews conducted with the general public, central and local government officials, politicians, businessmen, community leaders, private collectors (both formal and small scale), scavengers, and other individuals and institutions connected with waste management. Governance is taken here to be a relational concept emphasising the nature of interactions between the state and the social actors, and among the social actors themselves. At a sub-national level, urban governance is made up of a trinity of relationships between the central government and national institutions; local governments; and the civil society, including the private sector, NGOs and CBOs.

Urban Waste Management in a Governance Perspective

This chapter focuses on the management of solid waste, including the storage, collection, transfer, recycling, resource recovery and final disposal of waste. Solid waste includes refuse from households, non-hazardous solid (not sludge or semi-solid), waste from industrial and commercial establishments, refuse from institutions (including non-pathogenic waste from hospitals), market waste, yard waste, and street sweepings.

Dar es Salaam has a population of 3 million people, and is growing at over 7% per annum. It covers an area of some 1,400 km². Over 75% of the residents live in unplanned areas. Overall, the city is poorly endowed with infrastructure. It is managed by an appointed City Director, supervised by the elected Dar es Salaam City Council-DCC. The city is divided into three districts and 52 wards.[1]

Public Good Characteristics of Waste Management

Due to the high concentration of population and economic activities in urban areas, the waste generated cannot be disposed of efficiently on an individual basis. Once generated, it enters the public realm in that it can be disposed of on public or private land and can thus cause nuisances, environmental or health hazards affecting all society. Private households and firms often consider themselves exempt from any obligations after removing waste from their private domain. Waste management benefits all the community, in that any resident can enjoy the benefit of the service without diminishing the benefits for others. Thus, waste management is placed squarely within the public domain as a public good, and citizens expect the government to take actions and keep their environment clean.

The success of the authorities hinges on the availability of resources and good governance. Efficient management legitimises the state in the eyes of the public. Failure creates hostility and distances the public from the state, having important connotations for resource generation, democracy, transparency and accountability.

The Role of Urban Authorities in Waste Management

The Local Government (Urban Authorities) Act of 1982 lays considerable responsibility on urban authorities regarding waste collection and disposal. It requires that they remove refuse from any public or private place and provide and maintain public dustbins and other receptacles for the temporary deposit and collection of rubbish. The Act provides for the prevention and abatement of public nuisances which may be injurious to public health or to the good order of the area of the authority. At the same time, urban authorities are empowered to ensure that residents keep their premises and surroundings clean, in terms of the Township Rules made under the 1982 Act. The DDC has made a number of bylaws related to waste management, especially the 1993 Dar es Salaam City Council Bylaws (Disposal of Refuse).

According to the Township Rules, occupiers of any building are required to provide and properly maintain a receptacle for refuse storage. Garbage bins should be placed alongside roads for collection. The DDC has powers to require individuals to remove the accumulated refuse they have deposited anywhere, and throwing refuse on any street or public area is prohibited.

The Institutional Apparatus

Traditionally, solid waste management is the sole responsibility of urban authorities, although other higher levels of government have a major role to play. Given the failure of urban authorities to discharge their duties, central government has been forced to intervene now and again. Likewise, other actors, including the private sector, parastatal organisations, communities, small-scale business entrepreneurs and individual households have been involved in solid waste management.

Three DCC departments have responsibility for solid waste management, namely: the City and Social Welfare Department; the City Engineering Department; and the City Planning Department. Waste disposal is given little weight in the organisational set up of the DCC, being reduced to a section (Cleansing) of a sub-department (Preventive Services) of the Health and Social Welfare Department. Moreover, it is treated purely as a health issue. The Cleansing section is the executing body for waste collection and disposal, street sweeping and drain unblocking. It is also responsible for the formulation of policy in relation to solid waste management.

The activities of city cleansing are supervised by the Health Standing Committee, which plans, evaluates and advises on all matters concerning health, including waste removal and disposal. Viewing waste removal as a health issue subsumes other aspects of waste generation and management, and the concentration on cleansing does not bring to the forefront aspects such as waste generation and recycling. The private sector, CBOs and the public at large are not included in the waste management structure of the DCC.

The City Engineering Department is responsible for matters related to vehicles, plant and equipment, as well as purely engineering issues like roads leading to disposal sites. The Urban Planning Department is responsible for setting aside land for waste collection and disposal and it must liaise with the Ministry of Lands, Housing and Urban Development.

Any proposal from the Cleansing section passes through various stages before it can be approved. Starting from the section itself, it goes to the Health Department, then to the Health Standing Committee, on to the Finance and Management Committee, to the Full City Council, to the Regional Development Committee, to the Minister responsible for Local Government and finally to Parliament. This procedure can take a year or more to complete. Besides, the Cleansing section lacks autonomy even in crucial matters like the purchase of fuel or spare parts. It does not have a separate budget, and any money it collects goes to the DCC general revenue.

Solid Waste Generation and Handling in Dar es Salaam

Some 2,000 tons of solid waste are generated in Dar es Salaam daily as shown in Table 7.1. Despite various efforts to privatise solid waste collection, only about 10% of the waste generated is actually collected. Traditionally, it has been the DCC's responsibility to collect and dispose of solid waste. This duty used to be discharged successfully when the population was low, but the situation has slowly deteriorated since the late 1970s.

Table 7.1   Waste generation in Dar es Salaam, 1995

Waste Category	Amount of waste (Tons)	Share of total (%)
Domestic	935	46.8
Commercial	80	4.0
Institutional	185	9.3
Market	375	18.8
Industrial	225	11.2
Street cleaning	60	3.0
Car wrecks	40	2.0
Hazardous waste	50	2.5
Construction waste	15	0.8
Hospital waste	35	1.8
Total	**2,000**	**100.0**

Source: Dar Salaam City Council, 1995.

Some 800 people were employed in the Cleansing section of the DCC in 1994, and shortage of equipment and protective gear as well as low morale among the staff are major problems. Vehicles are inadequate, and the DCC cannot maintain them as required. Although steps have been undertaken time and again, usually with foreign or central government assistance, to procure new vehicles and equipment, they tend to get grounded after a short period due to lack of maintenance.

The DCC is also facing a problem concerning the final waste disposal site. Popular protest forced the DCC to stop using its long-time dumping site at Tabata in 1991. It was also barred from a number of other sites. Currently, waste is crudely dumped at a site in Vingunguti, which is a major environmental pollutant and nuisance to the area's low-income residents. There have been various protests against the continued use of

Vingunguti as a landfill site chiefly because the DCC has failed to improve the local environment and to operate the site as a sanitary landfill.

Environmental Hazards Associated with Improper Solid Waste Management

Lack of waste collection has forced people to dispose waste in a haphazard manner on common open spaces, by the road sides, and in ditches and drains, especially because there is not sufficient space to dig disposal pits on the sites they live. Besides being an eyesore and generating foul smells, waste haphazardly disposed encourages the breeding of disease transmitting flies, mosquitoes, rats and other vermin.

Unmanaged waste also blocks drains and causes seasonal flooding, leading to deleterious effects such as the damaging of roads. Leachate produced by the contact of waste with rainwater pollutes groundwater and river water. Scattered waste degrades the overall environment and leads to a decrease in land values. Uncollected waste leads to a decline in social mores, to increased scavenging, and erodes public authority. People often deal with waste by burning it, leading to further air pollution. Uncollected waste reflects badly on society as a whole, and may deter tourists, leading to a loss of foreign exchange.

Various surveys carried out in Dar es Salaam have indicated that solid waste is a major irritating problem to most residents; between 60% and 100% of those interviewed considered the DCC's solid waste collection services either bad or non-existent (Kironde, 1995).

Privatisation of Solid Waste Management in Dar es Salaam

Given such poor waste management conditions, the DCC, within the scope of the Sustainable Dar es Salaam Project (SDP) financed by UNDP and UNCHS, decided to embark on the privatisation of waste collection. In August 1992, the SDP organised a City Consultation on environmental issues, which identified waste management as a priority, recommending the immediate establishment of cross sectoral, multi-institutional working groups to implement a five point strategy of intervention, namely: an emergency clean up of the city; the privatisation of the collection system; the management of disposal sites; the establishment of community based collection systems; and support for waste recycling. Five corresponding working groups were set up.

Privatisation was made possible after the enactment of the 1993 By-Laws enabling the levying of Refuse Collection Charges (RCCs) from occupiers of residential, commercial or industrial premises. Failure to pay RCCs may lead to legal action, fines and/or imprisonment. The Bylaws also stipulated that no business license be issued unless RCCs have been paid. It was envisaged that, with the RCCs in place, solid waste management would be self financing.

Emergency Clean Up and Privatisation of Waste Collection

An emergency clean up of Dar es Salaam was undertaken in 1992-94. This short-term strategy aimed at clearing most of the heaps of refuse accumulated in different sites. The Prime Minister's Office, in collaboration with the donor community, raised US$1.4 million for this exercise, and this fund was used for refurbishing 30 garbage trucks, opening a new dump site and facilitating the DCC's day-to-day refuse collection services. Dump site management equipment was procured with the assistance of the Governments of Japan, Canada and Denmark. The two-year clean up was meant to prepare the way for privatisation. During this period, waste collection rose from 30 tons to 400 tons daily.

Privatisation is to take place in phases, with 10 central city wards and one contractor in Phase 1, and gradually extending the privatised areas to cover the whole city, with the participation of more contractors. Initial calculation showed that the contractor would be able to make a comfortable profit from the RCCs collected, despite the need to invest heavily in vehicles and equipment, to pay for the lease of DCC trucks and to meet waste disposal costs. In 1994, it was calculated that the annual cost to privately collect waste in the city centre was Tshs 641.8 million, while the revenue was calculated to be Tshs 678.6 million, thus leaving a profit of Tshs 36.8 million.[2]

Under the 1993 By-Laws, refuse collection charges were set so that rich areas could subsidise poor areas and commercial uses could subsidise residential uses. The DCC was supposed not to issue business licenses until RCCs had been paid a year in advance, and it was required to prosecute defaulters. Privatisation was supposed to generate resources to run the waste disposal site as a sanitary landfill, by requiring the collectors of solid waste to pay Tshs 800 for each ton disposed of at the landfill.

There is evidence that privatisation increased the amount of waste collected, from 30-60 tons daily collected by the DCC throughout the City, to 100-120 tons collected by the private contractor in the city centre alone. Resources were being used more efficiently. The private contractor in Phase 1 was able to collect around 100 tons of refuse per day with a labor force of 318 workers, while 800 DCC workers collect only between 30-60 tons. In the privatised area, collection went up initially to 75% of the generated refuse. Nevertheless, the private contractors' workers complained of poor wages, long working hours, lack of overtime, holidays or social security, etc.

Although at the beginning the contractor performed well, the amount collected subsequently decreased due to contractual problems which seriously affected revenue collection. This raised doubts about extending privatisation to the rest of the city, and the Phase 1 private contractor's area was reduced to 5 wards, from the original 10.

Despite the problems experienced in Phase 1, Phase 2 began in July 1996, with another four firms being commissioned to operate in another 13 wards. The performance of the DCC and the private contractors during the latter part of 1996 (cf. Table 7.2) indicates that, although both the private contractors and the DCC were doing well, their performance deteriorated towards the end of the year.

Table 7.2   **Average daily waste collection as recorded at the Vingunguti landfill site**

Waste collector/firm	Average daily refuse collection in the second half of 1996[a] (Tons)					Proportion of generated waste collected in December (%)
	Aug.	Sept.	Oct.	Nov.	Dec.	
DCC	180	168	164	106.2	120.9	16.1
Multinet (Phase I)	120	120	102	61.2[b]	42.4	31.5
Mazingira 1994 (Phase II)	-	40	40	15.5	11.2	2.7
Allyson Traders (Phase)	-	15	15	10.2	8.7	19.4
Kamp Enterprises (Phase II)	-	5	5	0.8	0.2	0.3
Kimangere (Phase II)	-	10	10	4.7	5.6	11.2
Total	**300**	**358**	**336**	**198.6**	**189**	**12.9**

a. Estimates based on the operational fleet and wards of operation.
b. Contractor withdrew from five of the ten wards originally contracted to be serviced.
Source: Mwihava (1997)

Problems in the Privatisation of Waste Collection

The City authorities have resorted to privatisation to tackle waste management in the city since 1995, but the experience so far is not encouraging. The private sector is not succeeding in collecting the waste or in collecting RCCs. The amount of waste collected remains low and contractors blame the low amount collected from the waste generators, which has made it difficult for them to invest in the necessary equipment. Poor waste collection leads to a vicious circle where the waste generators are discouraged from paying their RCCs.

An analysis of the situation reveals that the DCC itself was not well prepared for privatisation, perhaps because the idea was proposed by foreign experts, in a context where considerable resources were poured into the emergency clean up. It appears that the DCC's initial interest in privatisation was due to the large amounts of money expected to be obtained from businessmen and traders, as well as to the prospects of national and international financing. Because of this lack of preparation and the over-reliance on external advice and financing, the DCC is to blame for many of the problems in the privatisation exercise.

Originally, it was thought that the collection of RCCs would be easy since the 1993 By-Laws required that traders, businessmen and residential households paid them in advance. Unfortunately, none of these groups of would-be payers was consulted. The Ministry of Trade was not consulted either, and it eventually refused to support linking the approval of licenses to the payment of RCCs.

The determination of the value of RCCs was not related to the waste generated or to the cost of collecting it. Traders and businessmen who generated 33.4% of waste in the central areas were required to pay 80% of the total RCCs. Many of them showed that they could collect and dispose of their waste at a cost lower than what they were supposed to pay as RCCs. The payment of RCCs annually in advance was considered burdensome, since the sums of money involved were quite hefty. It was also thought unreasonable, since there was no past experience of efficient waste management by the DCC to refer to. Thus, it became increasingly difficult to convince the traders to commit huge funds in advance when they were not confident in the services and had little redress if these were not provided.

Many would-be RCCs payers refused to do so. The collection of RCCs in advance had been planned to be the key factor in enabling the contractors to invest in the necessary plant and equipment and offer incentives to workers. When the plan collapsed, the contractors were left to confront the businessmen, traders and household occupiers to get paid for waste collection. This proved difficult, as many refused to pay. A recent study involving two of the private contractors shows that the rate of RCCs collection was as low as 4.9% of the invoiced amount in one case, and 24% in the other (Table 7.3).

Moreover, the DCC failed to enforce the payment of RCCs and to prosecute defaulters, and the contractors were left marooned. Unable to force people to pay, they were left with limited resources, being unable to invest in equipment.

Table 7.3   Collection rate for two of the contractors, second half of 1996

Contractor	Invoice amount (Tshs)	Amount paid (Tshs)	RCC collection rate (%)
Multinet	169,397,200	40,699,731	24.0
Mazingira	50,932,250	2,499,500	4.9

Source: DCC (1997).

The DCC also failed to mount a public education campaign about the privatisation exercise, and the population was in the dark as to the whole experience, regarding the private waste collectors suspiciously as opportunists, particularly when they failed to deliver service to the satisfaction of customers. Besides, the general population had been used to free, albeit inefficient, refuse collection services and steps were necessary to convince them to start paying, particularly since the DCC was often accused of corruption. Payment of RCCs is relatively better in medium- to high- income areas compared to low-income areas, where the rate of RCCs is very low. This situation forced one of the contractors to withdraw his service from five low-income wards.

Lower ranks of the City authorities, community leaders, CBOs and NGOs were not effectively involved in this exercise either. A study carried out in 1995 found that between 90% and 100% of the population interviewed had never been consulted by DCC on the question of solid waste management in general, or of privatisation in particular (Kironde, 1995).

The relationship between the contractors and the DCC was often constrained. The DCC took long to operationalise various agreements required under the privatisation arrangements, without any apparent explanation. Proposals put forward by the contractors or by consultants to smoothen the exercise were usually not acted upon. The directory and supervisory role of the DCC appears to be wanting. It is doubtful whether a thorough evaluation of the tenderers was undertaken, particularly to ensure that they had the claimed resources. Monitoring the performance of the contractors proved to be inadequate and contractor obligations also proved difficult to enforce. The contractors did not appear to have the necessary resource at their disposal. All the same, without an assured stability of income from RCCs or other sources, they could not risk too much investment in the collection of waste.

As waste continued to accumulate in the city even in the privatised areas, the Prime Minister had to intervene in the DCC for six months to ensure that the city was clean. The DCC failed to do so, and the Prime Minister dissolved it and replaced it with a commission. The commission is still grappling with the problem of solid waste management and the privatisation option. Despite bringing in more contractors, both waste and RCCs collection have remained low.

Conclusion

Although the problems of the privatisation of solid waste management in Dar es Salaam may appear to be technical, they are essentially a problem of governance. Efficient waste management requires an adequate technical, financial and human resource capacity, as well as a reasonable degree of democracy, accountability and transparency.

Local authorities in Tanzania have long been under the control of central government, which has limited their scope of action by denying them competent man power, sufficient revenue, and a democratic organisation. Central government intervention has been in terms of *ad hoc* actions aimed at gaining political popularity, instead of long-term strategies directed to strengthening urban authorities. The result has been authorities that are not accountable or transparent to their electorate but that dependent on the central government and the international community.

Privatisation is viable option to dealing with solid waste in Dar es Salaam. So far however, success has been limited despite the engagement of five private contractors to collect waste in the City. A major constraint is the problem of who should pay the contractors. The arrangements put in place to enable privatisation expressed the local authorities' lack of accountability and transparency. If the general population is to be expected to pay for the services, it must be closely involved in the Council's plans and decisions. The tradition of inefficient urban authorities, with no good record of working closely with the populations and their organisations, discourages the growth of confidence among these people to pay for services.

Nearly three years of experimenting with privatisation have shown that the crucial matter to be confronted is the fact that the contractors cannot collect sufficient revenue from the RCCs from their areas of operation to be financially viable. An ongoing study on solid waste management suggests that the most important issues to be tackled are the need to reinforce the DCC's operational capability by the improvement of equipment and facilities and development of human resources; and the establishment of financial sources to increase the DCC's revenue generation (DCC, 1997). Privatisation could be a viable option when the contractors are paid by DCC for the services rendered, instead of the contractors having to collect the RCCs directly from the businessmen, traders and residential households. This requires the improvement in governance of Dar es Salaam.

Besides which, there are many actors that manage solid waste, working outside the recognition and involvement of the DCC. These include small scale operators, community groups, NGOs and CBOs. It is possible to slowly privatise waste collection and disposal by involving such actors in various forms of partnerships. The role of the DCC might be to identify and plan for disposal sites, to supervise the collectors and to enforce cleanliness regulations.

The main problem remains the continued isolation of the DCC from the general population and its belief that the solution to the environmental problems emanating from solid waste is to be found in the City Health Department. The improvement of solid waste management hinges in the overall improvement in the governance of Dar es Salaam, which largely means taking on board all the actors on the urban scene and operating in a transparent and accountable manner.

Notes

1   In June 1996, the Dar es Salaam City Council was dissolved for incompetence, and particularly for its failure to manage solid waste in the city. A City Commission was appointed to run the affairs of the city until a conductive

situation to restore the elected City Council was considered ripe. In this chapter, reference to the Dar es Salaam City Council should also be taken to refer to the Dar es Salaam City Commission when circumstances so require.

2   For reference purposes, the exchange rate in January 1997 was 596 Tshs to US$1.

References

DCC (1997), *The Study of the Solid Waste Management for the Dar es Salaam City*, Progress Report 3, JICA-Kokusai Kogyo Limited, March.

Kironde, J. M. Lusugga (1995), 'The Governance of Waste Management in African Cities: The Case of Dar es Salaam, Tanzania', report prepared for the *International Workshop on Waste Management in African Cities*, Ibadan, Nigeria, 4-6 September.

Mwihava, N.C.X. (1997), 'Lessons of Experience from the sustainable Dar es Salaam Project (SDP): Solid Waste Management', paper presented at the *National Consultation on the Replication of the Sustainable Cities Programme Nationwide*, Dar es Salaam, 12-13 February.

8 What's Health Got to Do with it? Using Environmental Health to Guide Priority-setting towards Equitable Environmental Management in Cities

CAROLYN STEPHENS

Introduction

The development of sustainable and equitable policy processes in metropolitan areas internationally is a major challenge. Few, if any, governments have taken this challenge on, even at the level of rhetoric. This is clear from the continued unsustainable and inequitable management of established and wealthy metropolis such as London or Washington, as much as for cities such as Accra, Sao Paulo or Calcutta.   It is also clear from the way in which ideas emerging from international agencies, such as Agenda 21 or Healthy Cities, with their explicit focus on peoples' quality of environment and life, are more often neatly subverted into the margins of strategic planning in cities, rather than becoming central issues for metropolitan development (Dooris, 1998).

This chapter aims to discuss briefly the ways in which environmental health criterion can be used to develop more sustainable and equitable urban policy. It will suggest that professionals of all disciplines must move together from rhetoric, against inertia, individualism and current trends, towards a real pursuit of urban equity and sustainability. A combined definition of such a goal would mean achieving adequate quality of life for all in the present generation, without compromising the needs of the future generations.

The policy background

It should be noted first that it is frankly hard to be optimistic about achievement of either equity or sustainability in the face of existing evidence. It is daunting to think that, by 2025, 3 out of 5 of us will live in urban areas. Most people will live in conditions which presage anything but health, and with options that resemble nothing like theoretical sustainability (UNCHS, 1996). Poverty internationally is concentrating in cities (Amis and Rakodi, 1994; Stephens et al, 1997). As 19th century analysts also noted, it is in the disaggregated health statistics of cities that we see the starkest implications of the continued maldistribution of social, environmental and political resources internationally.

Thus, this chapter is written at a time when the United Nations reports that the share of the poorest 20% of the world's people in global income stands at a miserable 1.1%, down from 1.4% in 1991 and 2.3% in 1960 (UNDP, 1997). At the same time, the value of combined assets of the 447 individuals who are billionaires exceeds the combined incomes of the poorest 50% of the world's population - some 1.3 billion people. These 447 individuals are the caricature and, through their visibility, the aspirational model, of over-consumption. It is even harder to maintain optimism in the light of evidence that current policy processes promote and are increasing, rather than lessening, inequities in the distribution and control of environmental, economic and political resources.

In 1997, even WHO, not noted for its commentary on policy processes, was led to observe that:

> Urban poverty has led in part to the introduction of economic reforms, but these very reforms often create new forms of poverty. These measures include shifting towards privatisation and limiting the ability of local government to sustain previous levels of services (WHOa, 1997).

Behind the statistics lie the dynamics of maldistribution between and within nations, within cities, and between individuals. This maldistribution of resources is reflected closely in patterns of health impacts - again, between and within nations, cities and individuals (UNDP, 1997).

Into this policy context comes the rhetoric of sustainable and equitable urban management. What would urban development priorities look like in real rather than rhetorical cities when they are based on criteria of equitable chances of life for the majority of people? The following sections address a small part of this overall conundrum, looking at the way in which environmental health criterion can be used to prioritise action and link different sectors. The chapter addresses principally the role of health data in promoting equity. This does not always mean that sustainability is promoted, but it can imply this, in re-directing over-consumption towards the poor, and in making planners consider long-term health impacts of urban policy. Following a discussion on the use of health criterion as advocacy for focusing policy on those with the least access to environmental conditions, I shall discuss the role of intersectoral coalitions in favour of equity.

Problem? What problem?

The policy process has to start with admitting that a problem exists, as how a problem is conceptualised and defined affects policy deeply. For urban policy this is particularly important. Until the 1980s, it was routinely assumed that, somehow, the process of urban development would automatically bring improved quality of life and health for a city's citizens. The links of urban environmental conditions and health in cities became clearer in recent years, in part due to more focused research (Bradley et al, 1992; Stephens et al, 1996; Songsore and McGranahan, 1998). Such research reveals that urban "health" is deeply compromised for a majority of urban people internationally, with little evidence that current strategies lead to improvements for the majority. Yet the routine use of health criterion as a key means of developing equitable urban environmental management is still to be explored. Indeed, urban economic policy overall is rarely based explicitly on people's health, despite the fact that the logical end of economic development is the improvement of peoples' quality of life and health (UNDP, 1994; 1997).

Health information on people's illness and death can re-focus debate on where urban policy is going. Simple descriptive work which documents accurately and in detail differences in environmental health impacts within cities can act as a lever to shift the policy process at the start point - recognising that a problem exists (Stephens et al, 1997; Akerman, 1996). Thus, health impacts related to water contamination can be linked to reliance of the poor on water vendors; can highlight the areas of a city where water contamination and epidemics occur most frequently; and can act as advocacy for use by public health professionals and planners

(Attipoe, 1997; Lewin, 1996; Cairncross and Kinnear, 1994).

Using health data as advocacy does not mean that urban policies will change in terms of agenda-setting or action, but it does mean that the "problem" is no longer invisible. This can catalyse debate and transparency. For example, in 1994, following a study of environmental health inequalities in Sao Paulo, a council woman stood for election using the research methods and results to sponsor a bill to routinely monitor and publicise inequalities in health impacts across the city, commenting:

> It is essential to have access to data and information to construct a just society. Studies put these issues in a very clear manner. I would like to draft a bill which enforces the municipal authorities to divulge data related to quality of life in a systematic way (Akerman, 1996).

Thus, at the most simple level, work which describes the unequal distribution of death and disease rates between groups within a city makes explicit any lack of logic in policies which argue that the "urban" context is inevitably healthy. These kinds of data do not change actions towards or within cities necessarily, but they facilitate transparency in the policy process. If city authorities disaggregate health data within cities routinely, they are able to reveal where health impacts in cities fall, and on whom they fall. Such data can be compiled by area in cities, revealing, for example, that the districts with the worst environmental and socio-economic conditions are also the districts with the worst health statistics (see Lewin et al, 1996; Stephens et al, 1994; Soton et al, 1995). If used strategically, this data enables planners within a city to begin to advocate and develop resource allocation mechanisms based on the city areas with the greatest problems. Again, this does not mean that policy actions will change towards more equitable resource allocation, but that resource allocation decisions become more transparently equitable or inequitable.

Descriptive data act as advocacy tools and make clear a policy process in which the rights of some groups to basic conditions supporting healthy life are not acknowledged. What they do not reveal is the other side of the coin: who bears the responsibility for the health situation? Who determines the policy process? Thus, most importantly, descriptions of urban health impacts cannot really suggest the next stage of the policy process: what to do? For this, one needs an understanding of the determinants of health impacts, whether they are individual or whether they lie within the processes which determines resource allocation in cities. One needs to understand the links between groups and individuals in cities in terms of resource consumption, resource allocation and health. One needs to broaden the debate, both to develop conceptual understanding and to gain a broader allegiance for policy change. The next section discusses this briefly.

Transparent problems - narrow actions?

Essentially, describing urban health impacts takes an urban policy debate as far as advocacy: it often places health of the majority, or the poorest, on the visible policy table of a decision-making minority in the urban society. It is often a technical argument posed to a political elite. This has always been so. In reality, the concepts of equity and sustainability within Agenda 21 only reiterate, with less eloquence, the ideas of writers from the 19th and early 20th century including Engels, Mumford and Kropotkin.

Most measures to ensure urban quality of life were developed from the thinkers in urban Europe, who worked to alleviate the conditions in the slums of industrial cities. The basic elements now seem simple and obvious. They do comprise, at least, a comprehensive set of physical interventions such as ample and potable water, sanitation, nutritious food, decent housing. However, and most importantly, historical gains were achieved within a political framework: the basic interventions were argued for through social and political process - they came with societal structures that placed education and enfranchisement on the agenda, and promoted remunerative, healthy work conditions for all - in societies which moved to share resources and decision-making processes equitably.

It important to highlight this point because it emphasises two issues: first, that 19th century health professionals worked in a time when health analyses were compiled as part of a multi-sectoral allegiance of professionals who argued for social justice (Acheson, 1992; Wing 1994). Thus, health specialists were part of a larger group within society which broadened their own debate about the causes of ill-health, and which broadened the constituency group arguing for change. Political scientists used health statistics, and health professionals argued for social change (e.g. Engels, Virchow, cited in Wing 1994). Second, advocacy was backed by a political process and wider decision-making facilitated by political scientists. Both were complementary and facilitated more equitable technical action.

Today's understanding of urban conditions and impacts does not often reflect the multi-sectoral policy breadth of the 19th century. For example, debates on health often take place largely within the health professionals' community in a city (Attipoe, 1997; Clauson-Kaas, 1997). This has two disadvantages, namely: it narrows the range of policy options in terms of both process and interventions, and it weakens the opportunity to shift policy through multi-sectoral and more participatory allegiances. The same could be said of environmental strategies (Clauson-Kaas, 1997; Biswas, 1997). In the majority of past urban environmental management strategies, attention has often been confined to the physical environment, namely land, water and air. Particular attention is often paid to the nature of water and air pollution and the activities of urban municipal services which directly bear on these two forms of pollution. Environmental management strategies emerging from such a perspective focus frequently on the provision and maintenance of physical infrastructure in order to improve the delivery of urban services, especially water supply, wastewater treatment and solid waste management, and in relation to the defined infrastructure, policy initiatives designed to improve the effectiveness of service delivery.

Many urban environmental management strategies are still, in effect, urban infrastructure development strategies. Yet a health-criterion based strategy must be founded on the recognition that the defining characteristic of the urban environment for the majority is poverty. Thus up to 60% of people in metropolises in so-called developing countries experience the environment as "poverty", namely severely impaired access to life sustaining and enhancing resources such as potable water, clean air, adequate shelter, decent employment and an acceptable income.

As Biswas's chapter in this book argued, an innovative strategy in Calcutta - the Calcutta Environmental Management Strategy and Action Plan (CEMSAP) - tried to address the issue of poverty centrally, and thus necessarily tried to go beyond improvements to the physical environment. Environmental problems are demonstrably related to the way in which an economy works and society is structured. The functioning of an economy is an immediate determinant of poverty, which is itself often directly associated with a bad and deteriorating environment. In CEMSAP it was recognised that, in order to mitigate many environmental problems, it was necessary to address underlying economic and social processes, which in Calcutta fundamentally revolve around issues of poverty and economic growth. CEMSAP encompassed the economic and social environments of

Calcutta as well as the city's physical environment.

A critical question from the start was how poverty could be addressed within an environmental management strategy. Poverty effects are most readily manifest in the manner in which the majority of the city's population experience their home and work environments, that is, as an unhealthy and difficult place in which to live and function. In these terms it was logical that within CEMSAP an environmental problem was primarily defined in public health terms as "risk" (e.g., polluted water, unsanitary living conditions, toxic industrial effluent, hazardous transport).

Disciplinary communication? Intersectoral collaboration?

One of the most valuable steps forward is broadening the conceptual debate towards poverty as a political disenfranchisement, and towards allegiances across sectors towards social justice. Work which links urban health inequalities to urban processes helps to open a larger discussion about health of people as a goal of urban policy. It can also begin to broaden a policy debate on urban inequality into a societal not a technical discussion. It opens a debate in which other professionals can participate, especially professionals from sectors which determine health more substantially: transport; planning; education; economic affairs; works and housing; water and sewerage.

For example, in CEMSAP, an analysis of environmental health impacts was used as a problem identification tool, complemented by a wide consultation process. An institutional, legislative and economic analysis was built around the problem identification to establish the processes which required change: identifying the process determinants of health. Action plans and policies were developed for different sectors, addressing the institutional processes and stressing complementary actions and keeping the goal of the strategy focused on health. Figure 8.1 draws on interdisciplinary frameworks being developed by UNEP and WHO and shows the ways in which different disciplines contribute at different levels in a policy arena.

This highlights a further point: most importantly, broadening a debate towards the process determinants of environmental health impacts in a city shifts policy towards discussion of the integrated system, not the actions of individuals in one area. It also shifts the overall policy discussion to one about political processes and institutional changes, shifting the debate from seeing impacts (a description) to understanding the reasons underlying the distribution for impacts, and importantly, the means out of inequity. It is possible, for example, using multiple analyses from different disciplinary and sectoral perspectives, to build up a conceptual picture of the process of inequity, not just the unequal outcome.

Figure 8.1

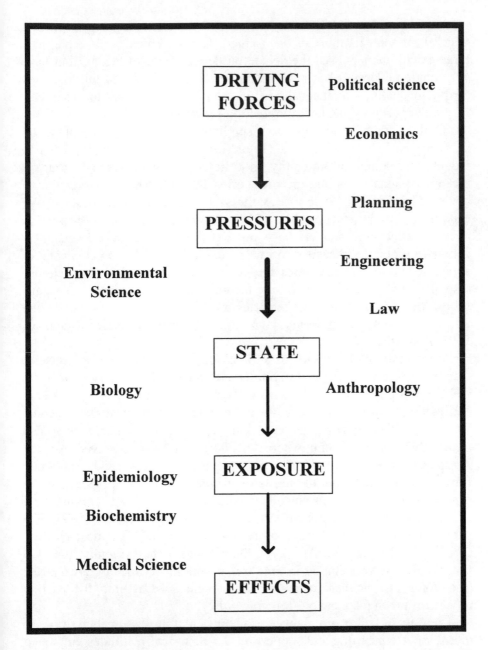

Source: Stephens (1998).

Conclusions: From rhetoric to mutual support

It has to be said that despite rhetoric, very little is shifting in the ways in which professionals think or are trained. Health professionals have, to a large extent, monopolised the debate on determinants of health, but have very limited conceptual or analytic understanding of how to link complex health inequalities to inequalities or inequities in access to resources, relative opportunity and decision-making processes.

Similarly, despite calls for interdisciplinary planning few environmental strategies include health specialists or form the types of coalition which would drive policy toward equity. In policy terms this is critical. It means that professionals working in cities, often all with the same overarching goals and beliefs, continue to box themselves into sectoral corners. We each have limited understanding of how to develop shifts in processes of resource allocation, and we have not developed skills and language to convince sectoral colleagues towards joint roles in the development of equitable and integrated actions across urban sectors.   We abnegate our responsibility and claim that we are technicians not politicians. Politicians follow this by claiming that their technical staff do not give them the information to use towards equity and against powerful lobbies (Stephens, 1998; Biswas, 1997).

At international level, there is a great deal of rhetoric, some of it useful clarion calls, much of it platitudinous and unrelated to reality. The 1996 UN Summit on Human Settlements in Istanbul placed goals of sustainable and equitable urban management on the international rhetorical agenda. The 1996 goals only articulate once again the principles which were voiced in the 1992 UN Rio Summit where Agenda 21 was first conceived. Agenda 21 places environmental concerns within a social and economic framework, starting from the needs of people: "Human beings are at the centre of concern for sustainable development.   They are entitled to a healthy and productive life in harmony with nature" (WHO, 1997). The parallel movement spearheaded by health professionals is that of Healthy Cities, and Health For All.   As Dooris has noted recently the two frameworks are very similar in time frames and theory (see Figure 2). He also notes that neither framework has found its way into the routine process of urban planning or economic policy.

There is, in reality, no point in academics and professionals writing on these issues concurring and supporting the rhetorical platitudes employed by the UN agencies. There is every need for us to recognise that the development of metropolitan areas currently does not bode well in terms of sustainability, health for the majority or equity. Academic integrity demands that we employ Gramsci's principles of "pessimism of the intellect and optimism of the will" in understanding (Gramsci,1996). The achievement of the latter in the current policy climate means that we have to re-assess and develop Kropotkin's principles that:

> ... in the practice of mutual aid ...we thus find the positive and undoubted origin of our ethical conceptions; and we can affirm that in the ethical progress of man, mutual support – not mutual struggle – has had the leading part. In its wide extension, even at the present time, we also see the best guarantee of a still loftier evolution of our race (Kropotkin, 1902).

Figure 8.2   Health for all and Agenda 21

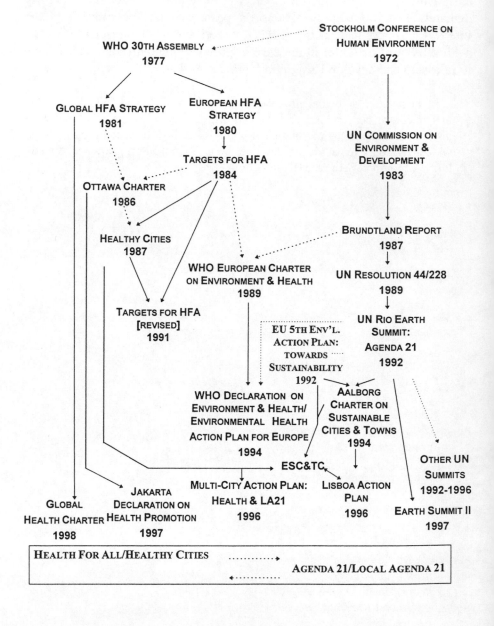

Source: Dooris (1998).

References

Acheson, D. (1992), 'The road to Rio. Paved with good intentions', *British Medical Journal*, vol. 304, May, pp. 1391-1392.

Akerman, M. (1996), 'Using composite indices to help capacity-building for intersectoral urban action', UNDP Marmaris Roundtable 'Cities for People in a Globalizing World', Office of Development Studies, UNDP.

Amis, P. and Rakodi, C. (1994), 'Urban poverty: Issues for research and policy', *Journal of International Development*, vol. 6, no. 5, pp. 627-634.

Attipoe, D. (1997), 'Dying for a change', *The International Health Exchange*, (August), pp. 8-10.

Biswas, K. (1997), 'Living in the city', *The International Health Exchange*, (August), pp. 6-8.

Bradley, D., Stephens, C., Harpham, T. and Cairncross, S. (1992), 'A Review of Environmental Health Impacts in Developing Country Cities', *Urban Management Program Discussion Paper no. 6*, The World Bank, Washington D.C.

Cairncross, S. and Kinnear, J. (1992), 'Elasticity of demand for water in Khartoum, Sudan', *Social Science & Medicine,* vol. 43, no. 2, pp. 183-189.

Clauson-Kaas, J. (1997), 'Chinese walls', *The International Health Exchange*, (August), pp. 10-12.

Government of West Bengal (1998), *Calcutta Environmental Management Strategy and Action Plan*, Department of Environment, Government of West Bengal.

Dooris, M. (1998), *UK National Roundtable Health and Local Agenda 21: Integrating Strategies in Local Government to achieve action in sustainable development*, United Nations Environment and Development UK Committee, Monograph of Create Centre, Bristol.

Engels, F. (1987), *The conditions of the working classes in England*, Penguin, London, 4th edition, pp. 1-293.

Gramsci, A., in Hoare, Q., and Nowell Smith, G. (eds.) (1996), *Selections from the prison notebooks of Antonio Gramsci*, Lawrence & Wishart, London.

Lewin, S. (1996), *Health and Environment Analysis for Decision-making: Case Study of Cape Town*, World Health Organisation, Office for Global and Integrated Environmental Health, Geneva/Medical Research Council of South Africa.

Schaefer, M. (1993), *Health, environment and development: Approaches to drafting country-level strategies for human well-being under Agenda 21*, World Health Organization Geneva, Switzerland.

Songsore, J., and McGranahan, G. (1998), 'The political Economy of Household Environmental Management: Gender, Environment and Epidemiology in the Greater Accra Metropolitan Area', *World Development*, vol. 26, no. 3, pp. 395-412.

Soton, A., Alihonou, E., Gandaho, T. and Dofonsou, M. (1995), *Health and Environment Analysis for Decision-making: Case Study of Cotonou*, World Health Organisation. Office for Global and Integrated Environmental Health, Geneva/CREDESA, Benin.

Stephens, C., Timaeus, I., Akerman, M., Avle, S., Borlina-Maia, P., Campanario, P., Doe, B., Lush, L., Tetteh, D. and Harpham, T. (1994), *Environment and Health in Developing Countries: An analysis of intra-urban differentials using existing data. Collaborative studies in Accra & Sao Paulo and analysis of urban data of four demographic and health surveys*, London School of Hygiene and Tropical Medicine, London, UK.

Stephens, C., Akerman M. and Borlina-Maia, P. (1995), 'Health and Environment in Sao Paulo, Brazil: methods of data linkage and questions for policy', *World Health Statistics Quarterly*, vol. 48, no. 295-108.

Stephens, C., McGranahan, G., Leonardi, G., et al (1996), 'Environmental health impacts', chapter in the UNEP/UNDP/World Bank/WRI 1996/97, *World Resources Report*, World Resources Institute, Washington D.C.

Stephens, C., Akerman, M., Doe, B., et al (1997), 'Urban Equity and Urban Health: Using existing data to understand inequalities in health and environment in Accra, Ghana and São Paulo, Brazil', *Environment and Urbanization*, vol. 9, no. 1, pp. 181-202.

Stephens, C., *Environment, Health and Development: Negotiating with complexity in priority-setting processes*, World Health Organization, Geneva/ Rockefeller Foundation (in press).

Stephens, C. (1998), *Environment, poverty and health in Accra, Ghana and Sao Paulo, Brazil: the response of the policy elite*, final report to the Economic and Social Research Council Global Environmental Change Programme, UK.

WHO (1997a), *Sustainable Development and Health; concepts, principles and framework for action for European cities and towns European Sustainable Development and Health series:1*, World Health Organization Regional Office for Europe, Copenhagen, Denmark.

WHO (1997b), *City planning for health and sustainable development*, European Sustainable Development and Health series 2, World Heath Organization Regional Office for Europe, Copenhagen, Denmark.

Wing, S. (1994), 'Limits of epidemiology', *Medicine & Global Survival*, vol. 1, no. 2, pp. 74-86.

United Nations Centre for Human Settlements (UNCHS) (1996), *Global Report on Human Settlements*, Oxford University Press.

United Nations Development Programme (UNDP) (1994), *Human Development Report 1994*, Oxford University Press, New York.

United Nations Development Programme (UNDP) (1997), *Human Development Report 1997*, Oxford University Press, New York.

PART II
MANAGEMENT

9 Management of the Urban Environment

MICHAEL MATTINGLY

Managing the urban environment is ultimately what this book is about. Policies and institutions are but elements of this management which uses them to achieve its ends. While this section groups a set of cases of particular interest from the standpoint of management, all cases collected in this volume add something to our knowledge of how environmental management is being approached and practised in urban areas at this time.

But what does it mean to manage the urban environment? The term management appears to be taken for granted in the environmental field, yet it needs identifiable substance if we are to know what to expect of it. A suitable definition can be constructed from the logic of its most common use in the world of business (Mattingly, 1994). This finds management to be: assuming responsibility for actions to achieve particular objectives with regard to a specific object. Using this model, the concern of this section is how responsibility is taken and exercised for actions to achieve an improvement in the quality of a city or town's environment.

The model focuses attention upon certain basic features when drawing lessons from the experiences collected here. These are a) the purposes of urban environmental management, b) the sense of responsibility involved, c) the actions undertaken and d) the object which is managed.

Discussions of management have spread from fields of development and urban affairs into environmental matters. They associate with urban environmental management a number of important secondary features. These include partnerships (UNCHS, 1997: 28; UNCHS, 1996: 310; Davey, 1993), participation (UNCHS, 1997: 27; UNCHS, 1996: 311), inter-sectoral approaches (UNCHS, 1997: 107; UNCHS 1996: 307; Davey, 1993: 8), and transparency and accountability (UNCHS, 1997: 107; Devas, 1993: 93; Davey, 1993: 30). These are also aspects of cases which it is useful to examine closely.

Responsibility

It is difficult to identify an adequate sense of responsibility behind the activities reported here. In the first place, there is much which seems to be 'business as usual': that is, traditional acts of public health and sanitation or of municipal engineering, or of housing policy implementation, acts which are not driven by a broader concern for the urban environment with its connotations of comprehensive scope, interrelated causes and effects and the **totality** of the effects. Problems tend to be tackled as if they are isolated and simple, and a realisation that complex coordinated action is required often seems to lie with observers (the authors) outside executing agencies. This is so with *favela* upgrading in Rio de Janerio (Edelman, Procee and Acioly), drain cleaning in Buenos Aires (Girardin and Greco), collecting rubbish in Colombo (Ratnayake) and in Copenhagen (Cooper), and the reduction of fuel emissions in Sao Paulo (Jacobi and Gouveia). These actions do not appear to be either generated or carried out because a responsibility has been felt to manage that city's environment, with its many dimensions of concern, interrelationships, and dynamics.

The Sustainable Programme for Accra (Doe and Tetteh) and the Urban Environmental Management Project in Thailand (Atkinson) are quite different. Both take a town or a city-wide perspective as a starting point before narrowing down to first actions. This approach has retained the complexity of the problems and their solutions, the multiplicity of the actions and actors, and a concern for the city's overall environmental state. There is very little else in this international trawl which depicts an attempt to activate responsibility across the usual organisational and jurisdictional boundaries. There is little else describing a programme at work with a responsibility as broad as the environment.

A second inadequacy is in the location of responsibility. Rarely do these accounts create the feeling that there are entities are taking charge of a city's environment (or try to). While certain of the fragmented efforts may be straining to assume larger roles (for example, the State Secretary for Environment for Sao Paulo wanting to move from control of emissions to restrictions on car usage, and perhaps then on to improving public transportation), they are a long way from wanting anything more than a bit of horizontal or vertical integration. Following their observation that the improvement of sewerage and drainage in the *favelas* can be linked to the pollution alleviation programmes for Rio's main water courses and the bay. Edelman et al comment that this is not part of city government's vision nor that of the citizenry.

Again, the Accra Sustainable Programme and the Thai UEMP offer practice which is considerably more advanced because it causes an institution to relate to the whole of the environment of the city or town. Priority actions are chosen because of their expected effects on the entire environment, not because of the competitiveness of a sector of spending, professional discipline or area of jurisdiction.

Purpose

In the absence of signs that overall responsibility for the environment has been taken up, one has to question the purposes behind much of the practice reported. Is not much of this practice no more than business as usual simply because actors remain in old habits of thinking, such that rubbish is to be collected, storm water is to be drained away, etc. A familiar aim of an operating department is not placed in a wider context and given greater meaning.

If this is too much to expect of established organisations, then indeed new ones are needed, as Girardin and Greco suggest for Buenos Aires, not only to take up responsibilities which ignore irrelevant boundaries, but also to establish an environmental improvement purpose behind which to coordinate, as best can be done, the efforts of a host of myopic actors. With regard to Sao Paulo, Jacobi and Gouveia point out the gains to be had through campaigns which raise awareness. To the extent that these influence leaders of existing institutions and their political masters, a redefinition of old purposes can be expected, as Atkinson finds in Thailand. Awareness raising conducted in Accra must lead to institutions adopting new purposes when they participate in a concerted effort to clean up the Korle Lagoon.

Actions

Most of the cases in this book are about actions. Actions are the fruit of management: doing as a consequence of thinking. On the one hand, there is evidence of world-wide efforts to counteract the deterioration of city environments which are certainly not confined to the More Developed Countries. And a full range of those conditions most likely to be grave is being addressed. Jacobi and Gouveia report reductions in industrial air pollution followed by changes in vehicles fuels to ones less polluting and then experiments with restrictions on vehicle use. Although it may not be as technologically advanced as Cooper describes in Copenhagen, waste management is being carried out in Colombo. Rio's *favelas* are being given service networks. Environmental management guidelines are being applied in Phuket and other urban places in Thailand.

One aim in grasping the urban management concept was to encourage new perspectives and innovation, to get institutions out of the rut of day to day repetition, if for no other reason than because that failed so badly. Yet, to some of these cases there is a feeling that routines are merely being repackaged as environmental management. However, Jacobi and Gouveia call attention to the audacity of banning all vehicles for a day in Sao Paulo and call for more of it if environmental management is to succeed. Copenhagen is innovating new methods of waste collection. Atkinson reports that the UEMP was created to pioneer new methods, and Doe and Tetteh claim that the Accra Sustainable Programme is there to do things differently.

On the other hand, when looked at in a harsh light, the accounts suggest that not nearly enough is going on, that they do not describe the tips of ice bergs but rather a thin crust on which to found hopes that the turmoil underneath can be contained. Not only do they often seem to be isolated (and not just by the authors' choices of boundaries) but, within their individual target areas, they tend to be inadequate without further steps. So Jacobi and Gouveia call for action to be extended to public transportation if air pollution is to be reduced; Ratnayake sees the current rubbish collection and disposal actions as rudimentary, and Cooper illustrates how, when they are taken much further, there is still more to be done; Edelman et al acknowledge that upgrading selected *favelas* will not be sufficient; and Girardin and Greco report how unblocking drainage in one part of the city creates more flooding elsewhere.

The issue here is not that of interconnectedness again, but that of scale. Is it that urban environmental management is only beginning to hold sway among decision makers? Or is lip service is being paid to it because of the pressures created by events like the Rio Conference? Or is it that engaging environmental matters at the level of a city is just too difficult and that the road ahead is too steep to be travelled swiftly? Local governments which are expected to manage urban areas are notoriously weak in capacity and resources outside of the most developed economies.

Object

Features of cities and towns are the objects at which the actions described in these papers are directed. The cases were chosen for this reason. On the threshold of the Century of the City, when the urban environment will surround most of the world's people as they work, play, and rest, it is logical the quality of this environment must be a major focus of environmental management in general. Yet cities and towns create new dimensions for both problems and solutions. Edelman et al assert moreover that as cities grow, their relationship to urban, national and global frameworks change significantly. Jurisdictions, actors, and processes are concentrated in towns and cities. Nevertheless, to mark out a geographic area of a particular scale as the stage for action - whether an urban or rural entity - leaves a great deal happening in the wings and even outside the theatre which can spoil the show. One looks in vain in these experiences for lessons which would clearly tell us if trying to manage the environment at the urban level is a mistake or a useful strategy.

At the urban level, it is clearly possible to engage the local population and businesses. This is a fundamental strength of the Accra Sustainable Programme and the Thai UEMP. The *favela* communities work with the city's consultants in Rio de Janeiro. As the literature on participation suggests (Moser, 1989, UNCHS, 1997), a commitment of additional local resources through greater sensitivity to local priorities, better identification of problems, and more political will seems to be forthcoming as a consequence. Perhaps there is also something at this scale which eases attempts to raise awareness, for it is hard to imagine how tools used in Accra and Thailand - working groups and town and city consultations - could operate meaningfully at a national or even regional level. And yet, as just noted, the actions taking place at this level seem so inadequate. Is this the result of making the urban environment the object to be managed?

Nevertheless, the shortcomings of an approach which is too local are noted elsewhere in this book, as are the difficulties of connecting urban and rural, city and region. The financial, technical, and organisational requirements of protecting and improving the environment may be well beyond the current capacities of institutions at the urban level. Ratnayake even sees an argument for a national approach to solid waste management in Sri Lanka.

Other Management Features

So this collection of cases suggests that in cities and towns around the world will be found attacks on individual environmental problems, yet organised efforts to manage an urban environment as a totality are possibly rare. Whether or not, in a strict sense, management of the urban environment is actually taking place, there is much evidence here of certain features currently associated with it. One could argue that these features are of greater importance anyway, and that interest in the management concept endures because it is a useful vehicle for selling such ideas as participation, partnership, an inter-sectoral approach, transparency and accountability.

Participation

Participation figures strongly in these accounts. Both the Accra Sustainability Programme and the UEMP in Thailand are built upon it, and it appears in various forms in practice - surveys of opinion regarding the Operaçao Rodizio in Sao Paulo, involvement of community associations in Rio's *Favela* Barrio Program - and its absence is a cause for concern in dealing with flooding in Buenos Aires. Participation is an effective means of raising awareness, without which there is not the political will or the individual motivation to protect or improve the environment. Participation is used or advocated as a means to motivate and organise the essential multiplicity of actors. It is used to obtain better information about conditions and potentials. It is employed to draw out priorities which have widespread support across departments and levels of government and among interests outside of government in the business and community sectors.

Partnership

A partnership is a form of participation. Some cases illustrate the mobilisation of business and household resources and capacities through partnerships. Partnerships between local governments and NGOs have been created in Thailand. Waste collection in Copenhagen is carried out by a non-profit cooperative company run by landlord and tenants associations, local government, and company representatives, but it was created 100 years ago. In many places it seems there is not the 'audacity' as in Sao Paulo to try a new approach. The Accra case bears watching as it lines up actors in the business and community sector behind the drive to reduce pollution of the Korle Lagoon. Will it, and the others of the Sustainable City Programme following this approach (UNCHS/UNEP, 1997a), succeed in capturing this way the resources required?

Inter-sectoral Approaches

Few of the situations described show that actions are taken across traditional divisions of responsibility (e.g. government levels, government departments), expertise (e.g. drainage engineering, squatter upgrading), or basic categories of actors (i.e. public, private and community). True, Copenhagen's waste management company brings together all three sectors, and fighting air pollution in Sao Paulo has progressed from enforcing regulations on industrial emissions to banning the use of cars. Moreover, the Accra Sustainable Programme and the UEMP in Thailand started much further along than this by recognising problems and solutions as complex and examining them with a host of the relevant stakeholders.

Nevertheless, if indeed most of today's claims of tackling environmental conditions in urban areas are little more than repackaged routines of public health and municipal engineering - such as rubbish collection in Colombo may be - it will be very difficult to achieve effective inter-sectoral approaches. The many attempts at awareness raising unfortunately are usually confined to the public served by a particular agency and are not turned upon the agency itself, as Ratnayake calls for in Colombo. The participatory features of the Thai UEMP and the Sustainable Programme in Accra show a way around this. They argue for similar centres of responsibility to manage the whole which can reveal to individual actors how their efforts mesh with one another and how they affect the overall situation with all its many facets and links.

Transparency and Accountability

It remains to comment on transparency and accountability which have been married by the drive for better governance, especially at the local level where government is expected to be more responsive. They are given little attention in these accounts of efforts to deal with environmental problems. Yet, by the logic of improved management, greater transparency in decision making and in the dealings of implementation will result from the participation of more actors of a greater diversity. It should follow from any step which heightens awareness of issues, problems, opportunities, and decisions to act, of who will perform what task, and of the agreed basis for their actions.

The point is to achieve more accountability, which will drive the quest for adequate environmental management and will press for efficiency and effectiveness in any management conducted. Here the difficulty of locating responsibility raises its head again and becomes a substantial challenge. For accountability to be a motivating force, it must settle on some entity (or entities). For it to be effective in improving the quality of government, accountability must fall upon an entity or entities capable of taking meaningful actions. Rarely does it seem that there is an institution which is genuinely taking responsibility for managing the environment of a particular city or town and which therefore can be held accountable. Moreover, where we find it, as in the Accra Sustainable Programme, it is an institution like an urban planning department, which when thoroughly tried will be found to be without the stature, jurisdiction, powers, skills, resources, and exclusivity of focus to do the job.

Conclusion

Despite the questions raised, may we conclude that management of the urban environment has become a widespread, if not yet major activity? After all, these accounts attest to a great deal that is going on at the urban level. Yet the matter of responsibility central to the concept of management used in this discussion gives cause to be dissatisfied. The cases provide few examples of a responsibility being assumed, much less being exercised, for a broad concern which is truly environmental. Instead, pieces of the environment are the objects of action, and there is weak evidence that the motivating purpose is to improve a piece because it is an important component of something larger which should be managed. Aside from the outstanding cases like that of Curitiba, Brazil (Rabinovitch with Leitmann, 1993), current accounts of urban environmental management actions - as opposed to intentions - tend toward similar conclusions (World Bank et al, 1996; Bartone et al, 1995; UNCHS/UNEP, 1997b; Gilbert et al, 1996).

References

Bartone, C., Bernstein, J., Leitmann, J. and Eigen, J. (1995), *Toward Environmental Strategies for Cities*, Urban Management Programme Policy Paper 18, World Bank, Washington, D. C.

Davey, K. (1993) *Managing Growing Cities*, Development Administration Group, University of Birmingham.

Devas, N. (1993), 'Evolving Approaches', in *Managing Fast Growing Cities*, Devas, N. and Rakodi, C. (eds), Longman, Harlow.

Gilbert, R., Stevenson, D., Girardet, H., and Stren, R. (1996), *Making Cities Work*, Earthscan Publications, London.

Mattingly, M. (1994), 'Meaning of Urban Management', *Cities*, Butterworth-Heinenmann, Oxford, vol. 11, No 3.

Moser, C. (1989), 'Community Participation in Urban Projects in the Third World', *Progress in Planning*, vol. 32, Part 2, pp 81-89, Pergamon, London.

Rabinovitch, J. with Leitmann, J. (1993), 'Environmental Innovation and Managment in Curitiba, Brazil', *Urban Management Programme Working Paper Series 1*, The World Bank, Washington D.C.

UNCHS (1996), *An Urbanizing World: Glzobal Report on Human Settlements*, Oxford University Press, Oxford.

UNCHS (1997), *The Istanbul Declaration and Habitat Agenda*, UNCHS, Nairobi.

UNCHS/UNEP (1997a), *Implementing the Urban Agenda*, UNCHS/UNEP, Nairobi.

UNCHS/UNEP (1997b), *City Experiences and International Support*, UNCHS/UNEP, Nairobi.

World Bank, OAS, IDB, and Inter-American Foundation, (1996), 'An Emerging Policy Agenda for Local Government: Final Report', the Second Inter-American Conference of Mayors, 17-19 April, Miami, Florida.

10 Air Pollution in São Paulo: The Challenge for Environmental Co-responsibility and Innovative Crisis Management

PEDRO JACOBI AND NELSON GOUVEIA

Introduction

Over the past several decades São Paulo had gone through a process of rapid population growth, industrialisation, urbanisation, and consequent metropolitanisation which has been accompanied by increasingly serious environmental problems. During the 1960s and 1970s São Paulo also experienced one of the fastest growth rates among cities of the developing world. If the current population of more than 10,1 million in 1995 is added to the inhabitants of the other 38 municipalities that compose this region, a total of 17 million inhabitants is reached, making it one of the three largest urban agglomerations of the world. It is the most wealthy city in Brazil, and its metropolitan area accounts for 18% of the country's GDP, 31% of the industrial domestic product and 25% of the industrial labour force (EMPLASA, 1994).

The city, as well as the metropolitan area, is characterised by great inequalities in income distribution, since the richest 10% of the population earns 30% of the total income, and the poorest 50% earn only one quarter of the total income (Rolnik et al, 1990). São Paulo's growth has created urban patterns similar to those in other Latin American cities, characterised by large disparities in health, social and economic conditions (Jacobi, 1990; Stephens, 1994; Akerman et al., 1995).

Although São Paulo is relatively well served by the so-called basic environmental urban services of water and sewerage supply and solid waste collection, their quality and quantity vary greatly between central, intermediate and suburban districts (Jacobi, 1995; Stephens, 1994; Akerman et al., 1995). The city of today is undergoing a severe environmental crisis as the result of persistently inadequate management by local authorities of the increasing and damaging problems linked to the following:

- chronic air pollution;
- a very serious delay in the expansion of the subway network and more adequate public transportation alternatives to enable a reduction in the use of cars;
- a very serious delay in the completion of the Sewerage System Expansion Plan;
- the constant reduction of green areas, which increases areas of flooding with severe social, economic and environmental impacts on the overall structure of the city;
- the contamination of most of the water sources and waterways within the city, and the risk that this involves for the population, mainly in the flooding areas; and
- the exhaustion of the conventional alternatives for the disposal of solid waste and the problems resulting from the contamination by waste of ground and surface water through runoff and leaching.

The Context of Risk: The Problem of Air Pollution[1]

The city of São Paulo faces specific problems of air pollution because of the combination of topography, climatic factors, and excess of motor vehicles. In the past few years air pollution levels have become a cause for great concern. While levels of Sulphur Dioxide (SO_2) have been declining with no violations of the World Health Organisation (WHO) standards (40-60 $\mu g/m^3$ for annual average), levels of fine Particulate Matter (PM_{10}) are gradually increasing after an evident decline up to 1989 (figure 10.1). Further, annual means of PM_{10} during this period were usually 20% to 70% higher than levels recommended by the Environmental Protection Agency (EPA-USA) for annual mean (50 $\mu g/m^3$).

Levels of Ozone (O_3) and Nitrogen Dioxide (NO_2) have remained fairly constant in the past few years. Despite this, guideline standards have been frequently exceeded for both on a daily basis. On the other hand, levels of Carbon Monoxide (CO) have been steadily increasing over the years.

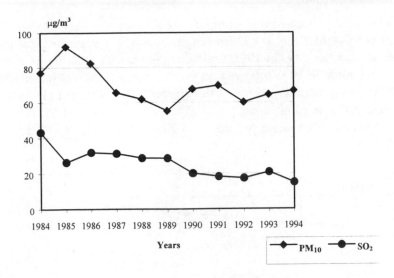

Figure 10.1 Annual mean levels ($\mu g/m^3$) of sulphur dioxide and inhalable particles in São Paulo, 1984-1994

Source: Gouveia (1997).

Therefore, despite several measures applied to improve the city's air quality in recent years, the air of São Paulo has been degraded by excessive levels of particulate matter, carbon monoxide, nitrogen dioxide and ozone which still exceed the Brazilian National Air Quality Standards (NAQS). These excesses occur every year in at least one or more monitoring sites by a factor of nearly 2 for PM_{10} and Ozone and 1.5 for Oxides of Nitrogen.

Initially associated with industrial production, which has now reduced its impact significantly, air pollution nowadays is mostly produced by motor vehicles and related activities, which are in general responsible for 40% to 90% of the pollutants present in the air (CETESB, 1995). The vehicle fleet in the city of São Paulo is estimated in 4.5 million cars and 12,000 buses. This fleet has been expanding rapidly in the past few years. While population growth had stabilised between 1980 and the early 1990s, and the urbanisation process had slowed down, the vehicle fleet expanded from 1,6 to 3,8 million vehicles in the same period (figure 10.2), an increase of 140%.

The Traffic Engineering Company (CET) of the city of São Paulo measures systematically the mean length of traffic jams in several critical points around the city. In 1992, traffic congestion at peak times averaged 36 km in length; in 1995 this grew to 94 km and in 1996 it reached 190 km, becoming the second most serious problem for the population, after violence. The rate of occupancy is 1,5 persons per car, and the number of cars estimated to circulate per day is 3,200,000 (CETESB, 1995; Sobral, 1995).

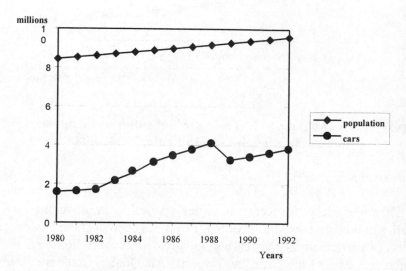

Figure 10.2 Trends in population and number of motor vehicles (in millions) in São Paulo, 1980-1992

Note: due to changes in registration of all vehicles from 1988 to 1990 there was an artificial decrease in the fleet during this period.

Source: CET (1992).

Consequently, there has been an expansion of the so called critical areas of air pollution and air quality has significantly deteriorated. During the first 5 months of 1997, air quality standards were exceeded on 40% of the days, on average a polluted day every other day (*Folha de São Paulo*, 1997).

Impacts

The biological mechanisms through which air pollutants can produce adverse health effects are mostly based on laboratory experiments. Most of the effects are related to the respiratory system since it is the first target once air pollutants are inhaled. Gaseous air pollutants exert a direct irritant effect on the airways. For particles, the mechanisms are related to the size and the water solubility. Both determine the capacity of the particles to reach deepest parts of the respiratory system. In summary, air pollutants can have a direct action on the lungs' parenchyma leading to chronic obstructive airway disease, irritation and broncho-constriction and an indirect action making the lungs more susceptible to infections (CETESB, 1993).

The escalating deterioration of air quality in the city of São Paulo has made air pollution one of the biggest causes for concern for the majority of the population. A risk-perception household survey identified air pollution as one of the three major environmental problems afflicting people's lives, together with water and poverty (Jacobi, 1995). Its major impact seems to be on the health of exposed individuals and yet, due to high episodes in the past few years, health warnings were frequently issued for the whole of the metropolitan area (Oliveira and Leitmann, 1994). Studies have shown that levels of air pollution in São Paulo are high enough to produce adverse health effects such as increases in mortality and hospital admissions. It is estimated that increases in daily levels of some of the most common urban air pollutants are associated with 5% to 8% increase in deaths in the elderly from respiratory and cardio-vascular diseases, and 13% to 30% increase in hospitalisation from respiratory diseases in children.

For mortality it seems that the main victims are the elderly and persons who suffer from chronic respiratory diseases such as asthma and bronchitis. The risk is much higher in those groups. However, children suffering from malnutrition might also be at a higher risk of succumbing to such exposure (Saldiva, 1995/96). Data provided by one of the largest specialised paediatric health institution in São Paulo indicated that there was an increase of 20% in the number of hospitalisations due to respiratory problems for the months of May to September, when the quality of air suffers significant deterioration (CEDEC, 1995/96). It should also be stressed that pneumonia infections are the main cause of mortality for children under five years of age in the city of São Paulo.

Operational Solutions and Management

With industrialisation in the 1960s and 1970s the problem of air pollution in São Paulo soon became a concern for most citizens and government and it was introduced into the political agenda. Emission control was started in 1973 by the Company of Environmental Technology and Sanitation (CETESB) of the State Secretary for Environment and it has been relatively successful in reducing overall levels of industrial air pollution, especially during the period of industrial growth. By enforcing emission standards, penalising the industries when their emissions were above permitted levels, and mandating use of best practice for the largest sources of industrial pollution, CETESB has achieved large reductions in emissions of SO_2, CO, and NO_2 (CETESB, 1993).

Control of emissions from mobile sources has also been carried out by CETESB. Since 1976 there has been a control programme on emissions of black smoke from heavy duty diesel vehicles and since 1986 the National Environmental Council enacted a resolution establishing the automotive emission control program nation-wide under the name PROCONVE. This Program was made possible through agreements with the automobile industry that incorporated in its projects advanced technology (electronic injection and catalyser) to help reduce emissions of pollutants. In 1992 the new legislation required catalysers in new cars and emissions of a maximum of 12 g/km, progressively enabling a significant reduction of automobile emissions, but far behind the requirements of cities of the first world. For 1997 the acceptable limit of emission is set to be of 2.0g/km.

Another important measure adopted was the addition of alcohol to petrol instead of lead. This produced a reduction in the emissions of CO, keeping hydrocarbons and NO_2 at the same levels. In addition, this modification in the petrol mixture has resulted in a considerable reduction in lead levels in São Paulo (CETESB, 1993).

Moreover, during the 1980s a substantial change occurred in the vehicle fleet with the introduction of alcohol engine cars that gradually replaced the petrol engine ones. In 1980 the first alcohol engine cars were introduced and by 1989 they accounted for 49% of the fleet. The petrol mixed with 22% of alcohol and the alcohol itself are two fuels with low pollutant potential. These were pioneering measures, devised and introduced in Brazil during the early 1980s and represented a success in the control of vehicle emissions comparable with the USA and Europe but in half the time (CETESB, 1993).

However, despite the fact that half of the cars run on less-polluting alcohol, currently, there are an estimated 4.5 million motorised vehicles in SP, a fleet which is growing at the rate of 5-10% annually (Oliveira and Leitman, 1994; CET, 1992) The efforts for emission reduction have been offset by the huge increase in the number of vehicles, often accompanied by badly maintained and improperly controlled diesel engines. In addition, the ever-growing traffic congestion and reduction of the average speed of the cars increases the emissions of each individual vehicle. This is compounded by insufficient and inadequate provision of public transport, encouraging the use of individual transportation.

Policies and Programmes

In recent years, as concern was growing about levels of air pollution, public authorities have taken differing measures to tackle this problem. The first action happened in 1988, when air pollution reached very high levels. Public authorities declared an emergency interdiction of the central area of the city for the circulation of vehicles during one day. This was an experimental attempt to reduce the high levels of air pollution produced by car emissions. Since then, other attempts have been made to restrict the circulation of vehicles in central areas of the city, however none of them were regulated (CEDEC, 1995/96).

In 1995, the environmental agencies introduced in their agenda some of the main control mechanisms practised in other great metropolis such as Mexico City and Santiago de Chile. These consisted of 'rodizio' (systematically prohibit the circulation of vehicles once a week, according to the registration plate/number) and the compulsory inspection of vehicle emissions. The federal and state governments are currently discussing the implementation of a Program of Inspection and Maintenance of Vehicles which foresees the compulsory inspection of the fleet in circulation through authorised stations in cities that have critical air pollution situations.

Non-conventional Air Pollution Prevention Policy

As a preventive measure the State Secretariat of the Environment has implemented, since 1995, the 'Operação Rodizio' to operate during the winter months, when pollution levels increase. The rationale for this procedure was the need to remove a daily percentage of cars in circulation, reducing levels of emission and traffic congestion. From its beginning the program was presented as an emergency and civil defence measure. It was an emergency because with the proximity of the winter season, public authorities felt that an urgent measure had to be taken.

During one week in 1995, an experimental and non-compulsory program was implemented and it divided opinions between specialists and society in general. For the critics it barely represented a mitigating measure, and given the precariousness of the public transportation system, the imposed costs to society would not compensate for the eventual benefits. Those in favour considered that it was successful, for it achieved the support of 38% of the drivers during the whole week, which was good considering that it was voluntary, without the application of any sanction, that it was poorly advertised, and that it had no support from the municipal authorities. In spite of its short duration, the impacts on air quality were felt, namely a reduction of carbon monoxide levels.

The success of this experience encouraged the Environment Secretariat to propose implementation of the scheme on a compulsory basis for the months of June-August 1996. However, before its implementation, this programme had to be voted for by the State Assembly to be transformed into a law. After intense negotiations, the law was passed but the time period was reduced only to the month of August (7am to 8pm), and buses, trucks and school buses were excluded. From this moment the 'Rodizio' was no longer a voluntary activity but a compulsory one with a U$100 penalty for non-compliance.

The main obstacles to greater success arise from problems of operating the program and from the continuing bad quality of public transportation. The existing subway network, with only three lines extending 44 km, and with almost no investment in the last decade, is inadequate for future city expansion.

Impacts

Several initiatives were taken by those opposing the Program, but none of them succeeded, and the 'Operação Rodizio' was implemented. Various activities were developed in the preceding weeks such as distribution of pamphlets in streets and schools and public debates to stimulate discussion. The results presented in official reports from the Secretariat were relevant and indicated the success of this complex initiative.

The average compliance during the whole month of August, 1996 was 95%, representing a withdrawal of 456,000 cars/day, and a reduction of 329 ton/day of CO. The relative emission of carbon monoxide of cars fell from 66% to 51%. The effect of the average fleet withdrawn, plus that of the greater fluidity of the circulating fleet, represented a total reduction of CO of 529.3 ton/day. There was an increase from 16km/h to 20km/h in the average speed of buses, and this produced an increase of 2% in the number of daily trips. The average reduction of traffic congestion was 39%. During the month nearly 170,000 fines were charged (SEMA, 1996).

After the completion of 'Operação Rodizio' public opinion was consulted, and research conducted by one of the most influent newspapers in the country indicated that 58% of the population considered that this program had to continue.

Conclusions

Most of the policies and programmes practised so far to reduce the levels of air pollution in São Paulo have focused on making vehicles cleaner. While car-related emissions have been shown to be the most serious threat to air quality in São Paulo such approaches have been palliatives and rapidly counterbalanced by the continuous increase in the vehicle fleet.

Urban air pollution, as well as many other environmental problems of urban areas, is a result of a complexity of factors which are most times related to the process of urbanisation itself. It seems that for public authorities, it has been more important, and to a certain extent more convenient, to use selective approaches which focus on improving the economic development and the productive efficiency of a city. Minimum concern or none at all has been given to more comprehensive approaches which include the actual environment in which the population will be living and which, therefore, will influence the quality of life.

The problem of air pollution in São Paulo is complex with multiple determinants. In order to reduce the damaging levels of air pollution, one needs to target not only car-emission reduction but also the provision of mass transport of good quality and quantity, the alleviation of traffic congestion, and reduction in the use of private cars. Equally important is the need to promote education and research. This implies that environment, health, education, economy, transport and urban planning sectors are all interrelated.

Further understanding of the inter-relation between these various matters is fundamental, and it is particularly important to break out of the narrow framework within which each is considered. Without some intersectoral planning, any intervention will be of limited use in substantially reducing air pollution. Moreover, further research is needed to assess the exact public health impact of urban air pollution in different environments, climates and socio-economic contexts. Similarly, the long-term effectiveness of current and past policies in terms of social and health benefits should be evaluated.

The successful experience of Operação Rodizio, although it had limitations, indicates that there is a need for the public authorities to be innovative and audacious in formulating and implementing environmentally concerned public policies. The program can be seen as an environmental education experience on a large scale, where a process using penalties built up step by step a practice that affects vested interests. It was an excellent opportunity to publicly debate the crisis management of environmental degradation in megacities.

What has to be stressed is that implementation of non-conventional policies and programs to prevent environmental degradation is very complex, and that in the recent years only this initiative (Operação Rodizio) can be considered innovative because of the institutional engineering on which it is based. The main elements of this initiative link the need to increase people's information about the existing environmental risks and the importance of their input of social capital (willingness to participate in the positive outcomes of public policies) as part of a necessary democratic interaction between local government and citizens.

The implementation implies not only socio-political articulation, but also agreement with the basic idea that a process is not only proposed but made public. In addition, the process entails dissemination through public information campaigns and consultative mechanisms oriented towards building the capacity of the community to stimulate and consolidate an efficient and consistent process of participation.

The burden of exposure to air pollution is an almost inevitable feature of urban living throughout the world. Moreover, the share of this burden is usually unequal since the majority experiences the problem while the minority, the car owners, creates it. As with other environmental problems, the worst of the air pollution usually remains with the politically weaker and more socially deprived groups. If these different sectors of society can identify what they have in common and can work effectively together, they will have the possibility of a very powerful political force.

Note

1 Unless otherwise indicated, the information in this section is taken from Gouveia (1997).

References

Akerman, M., Campanario, P., and Maia, P. B. (1996), 'Saúde e meio ambiente: análise dos diferenciais intra-urbanos, Município de São Paulo, Brasil', *Rev. Saúde Pública*, vol. 30, no. 4, pp. 372-382.

CEDEC (1995/96), 'Poluição Atmosférica', in *Debates Sócio-Ambientais 2*, Centro de Estudos de Cultura Contemporânea, São Paulo.

CET (1992), 'Acidentes de trânsito - 1992', Companhia de Engenharia de Tráfego, São Paulo.

CETESB (1993), 'Relatório de Qualidade do Ar no Estado de São Paulo-1992', Companhia de Tecnologia de Saneamento Ambiental, São Paulo.

CETESB (1995), 'Relatório de Qualidade do Ar no Estado de São Paulo-1995', Companhia de Tecnologia de Saneamento Ambiental, São Paulo.

EMPLASA (1994), 'Sumário de Dados da Grande São Paulo', Secretaria de Planejamento do Estado de São Paulo, São Paulo.

Folha de São Paulo (1997), '1997 Teve 40% do Dias Poluídos em SP', Folha de São Paulo 05/06/1997.

Gouveia, N. (1997), 'Air Pollution and Health Effects in Sao Paulo, Brazil: A Time Series Analysis', PhD Thesis, University of London

Jacobi, P. (1990), 'Habitat and health in the municipality of São Paulo', in *Environment and Urbanization*, vol. 2, no. 2, pp. 33-45, IIED, London.

Jacobi, P. (1995), 'Environmental Problems Facing Urban Households in the City of São Paulo, Brazil', Stockholm Environment Institute, Stockholm.

Oliveira, C. and Leitmann, J. (1994), 'Urban Environmental Profile: São Paulo', *Cities*, vol. 11, no. 1, pp. 10-14.

Rolnik, R., Kowarik, L., and Somekh, N. (1990), 'São Paulo, Crise e Mudança', Editora Brasiliense, São Paulo.

Saldiva, P. (1995/96), 'Efeitos da Poluição Atmosférica na Saúde', in *Debates Sócio-Ambientais*, vol. 3, no. 26, Centro de Estudos de Cultura Contemporânea, São Paulo.

SEMA (1996), 'Balanço da Operação Respira São Paulo', Secretaria do Meio Ambiente, São Paulo.

Sobral, H. (1995), 'O Meio Ambiente e a Cidade de São Paulo', Makron Books, São Paulo.

Stephens, C. et al (1994), 'Environment and Health in Developing Countries: an Analysis of Intra-Urban Differentials Using Existing Data', Monograph, London School of Hygiene and Tropical Medicine.

11 Sustainable Urban Development and the Urban Poor in Rio de Janeiro

DAVID J. EDELMAN, PAUL PROCEE AND CLAUDIO ACIOLY JR.

Introduction

Environmental degradation resulting from cities is not primarily a result of either the urbanisation process or a shortage of environmental resources such as land and fresh water. It is rather caused by economic and political factors. In many cases, local governments lack the means and instruments to manage the process of urban development effectively. Poor governance is at the root of the resulting problems, which include failures to control industrial pollution, to provide basic sanitation services to city-dwellers, to ensure that sufficient land is made available for housing development, to generate resources and investments and mobilise the participation of key actors, to maintain green and recreational areas and to enable the appropriate disposal of wastes. This ineffective governance is often linked to economic weakness and an unstable political situation characterised by a short-term planning perspective and, moreover, corruption.

Urban Environmental Management, Cities and Sustainable Development

Urban environmental management is best seen as a subsidiary process taking place within the overall process of urban management. It is an all embracing concept covering not only the physical environment, but also issues related to the urban economy, incomes, infrastructure, investments and institutions, all of which must be seen in relation to the political, social and cultural environment of any urban area. Conceptually, all basic principles of urban management also hold for urban environmental management.

The density of many different kinds of human activities taking place within a relatively small location, attracting resources from far beyond the administrative borders of the city, has obvious advantages or potential advantages for meeting the goals of sustainable development (Mitlin and Satterthwaite, 1994) and tends to maximise the benefits of economies of scale. These include:

- effectively responding to social and health needs;
- minimising the use or waste of non-renewable resources;
- sustainable use of renewable resources; and
- keeping wastes generated by city producers and consumers within the absorptive capacity of local and global sinks.

However in the absence of effective urban governance, the advantages and potential advantages pointed out previously can be transformed into enormous problems, especially when local governments are not well-equipped to deal with the complexity of urban development. The high concentration of industries near households can generate health problems that can be easily be transmitted to a large number of people. Problems with water management and flooding can be substantial since the water from large catchment areas flows into single streams whose volume can increase tremendously and cause disasters in and around cities. In addition, poor and illegal settlements are often constructed in flood plains and on steep slopes in many cities of developing countries as a result of the absence of a planning framework. Inadequate housing supplies can induce informal urbanisation and overcrowding, leading to the occupation of land unsuitable for human settlements. Thus, for these and many other reasons, cities are indeed hazardous to the world's ecological system if not properly managed.

Urban Governance in Brazil

Brazil's municipalities have enjoyed a significant level of autonomy since 1934 and which have the power to adopt laws, to levy and collect taxes, to organise their administrations, to define budgets and priority areas for investment, to conduct of urban planning and provide housing and urban services to their inhabitants and to oversee the welfare of the population. Municipalities also have the right to formulate and approve their own organic laws, which act as municipal constitutions, as long as they do not conflict with the federal and state constitutions. Furthermore, the mayor, vice-mayor and the councillors of municipal legislatures are democratically elected. Aside from the period of military rule (1964-85), this level of responsibility and authority places cities in Brazil among the most powerful of those in developing countries vis-à-vis their national governments.

In 1996, local governments invested 35% of the total financial resources of the public sector nation-wide, were responsible for 25% of the consumption generated by the governments and accounted for 17% of the total expenditures of the three levels of government (JB, 1996). Since the 1988 constitution was enacted, Brazil's cities have had unprecedented revenues and have spent heavily in the social sector. As a consequence, a number of innovative forms of municipal management and participatory planning have emerged in Brazil. Some valuable experiences are the participatory budgeting in Porto Alegre, Brasilia, Vitoria and Belo Horizonte; the popular councils in Vitoria and Fortaleza; the management of solid waste and recycling with "social fares" in public transport in Curitiba; the self-management of mutual aid housing in São Paulo, and the paid mutual aid and reforestation programme in Rio de Janeiro, just to name a few.

The Municipality of Rio de Janeiro

While Rio de Janeiro with a metropolitan area population of more than 10 million and a GDP which is twice that of Egypt's is a thoroughly modern and professionally managed city known for its natural beauty, it is, at the same time, well known for its *favelas* or squatter settlements. During the eighties and beginning of the nineties Brazil suffered a severe recession. The housing situation deteriorated and the *favelas* and clandestine settlements - illegal land subdivisions undertaken by private developers in peripheral areas - mushroomed and rapidly increased in density. In 1991, 33% of the registered housing units in the municipality were located in *favelas*, illegal land subdivisions and low income public housing estates, providing accommodation to more than 2 million inhabitants (IPLANRIO, 1993).

In addition, high inflation and disparities of income during the recession aggravated social tension and economic insecurity, and they resulted in a sudden and uncontrolled increase of violence and criminality never before experienced before in the city. Local government administration was absent in *favelas*, and the needs of the urban poor were neglected. The *favelas* were gradually taken over by outlaws and criminal organisations involved in drugs and weapons and various other illegal activities. In the absence of the State and the law, parallel and informal structures of power were established in several low income settlements by criminal organisations. This phenomenon was exacerbated by a weak and often corrupt police apparatus. The residents' associations - so active during the 1970s and 1980s - were intimidated and lost influence; sometimes people were murdered if they did not co-operate.

In addition to these socio-political difficulties, Rio's urban environment is characterised by a unique and vulnerable topography. Its 86 km of coastline are dominated by mountain ranges and massifs which delineate specific limits for the areas suitable for human settlement. A severe and long lasting housing shortage, coupled with land speculation, have pushed human settlements towards extremely vulnerable sites, which represent the only options left for the poor. The metropolitan area's 926 squatter settlements and illegal subdivisions have spread throughout the region according to IPLANRIO (1993); and many have been built on vacant land, subject to flooding, and on sloping areas up to 100 meters above sea level. The process of deforestation in these areas has further affected the retention capacity of the soil and the stability of slopes, and the increased erosion has resulted in sedimentation of rivers and streams.

Restoring Local Governance in Rio

The previous administration in Rio (1993-96) took office with a determined political will to put an end on what it called the urban chaos and the absence of governance. The municipality was restructured to ensure a more efficient management system. A new financial management procedure was introduced in order to be more responsive to externalities and take rapid decisions. Investments in public security, modernisation of the police apparatus and the development of social programmes gained prominent positions in the political agenda. A strategic planning process was launched and provided different mechanisms at different levels for participation of the stakeholders in the formulation of what is probably the first strategic urban plan formulated in Latin America. This helped the municipality to design a realistic plan taking into account the opportunities and threats for the development of the city.

Furthermore, the urban property cadaster was computerised and linked to fiscal and budgetary management. Project teams were set up to manage specific sectoral plans and an urban reform initiative was launched through the execution of several strategic programmes and projects targeting the informal and the formal city. The programme that deals with informal settlement upgrading deserves particular attention here since it highlights the urban environmental improvement agenda for the city's *favelas*.

An extremely interesting point here is that this programme was initially self-financed by the municipality. Rio is one of the few large Brazilian cities with an up-to-date cadaster and a geographic reference database which is continuously processed and improved by IPLANRIO. It provides a solid source of revenue from municipal property taxes, reaching US$ 770 million in 1992. By the end of 1995, the municipality had managed to build up a reserve of US$ 1 billion for investments. It issued municipal bonds and sold them on the international market, generating a small surplus of US$ 125 million to be used as well to finance its investment plans.

The availability of finance allowed the municipality to launch two important urban reform programmes. The first, the Rio Cidade Programme, was intended to renew 19 neighbourhoods and recapture the city for its citizens, while the second, the *Favela Bairro* Programme, was formulated for informal settlement upgrading. The latter programme is the focus of the following analysis.

The Profile of *Favelas*

Although the profile of *favelas* is widely known, it is worth mentioning a few characteristics of the areas in which the urban poor live in Rio. In 1991, the census registered 412 *favelas*. After IPLANRIO's survey in 1993, 570 settlements were registered with an estimated 1.3 million inhabitants. The settlements are characterised by the precarious provision or total absence of urban infrastructure and public services. The access roads are very narrow and hardly follow any alignment; the plots are irregular in form and size, and there is rarely any open space for leisure. There is no tenure formalisation either, except where the local government is carrying out upgrading and formalisation projects. The dwellings are built with wood or ceramic bricks, but the percentage of wood-based dwellings has decreased rapidly over the last 10 years.

During this period, *favelas* have also been consolidated and their housing stock transformed, and this was tolerated by the local government. However, many *favelas* increased in density over the last 10 years by building upward, while others expanded onto steep hills and higher areas. This hindered accessibility, increased the risks of landslides and aggravated other local environmental conditions. The deforestation and occupation of steep areas also caused severe environmental impacts at the city level, with heavy rains in 1995 causing substantial material damage and loss of life in the city through flooding.

A New Housing Policy Framework

In 1994, the municipal government established the Municipal Housing Secretariat (SMH) with the specific mission to formulate, co-ordinate and execute municipal housing policy. Several initiatives and projects where launched to achieve a sustainable solution for the housing shortage in the municipality. In order to cope with the acute housing problem, the SMH created the Upgrading of Popular Human Settlements Programme (PROAP/RIO), also called the *Favela Bairro* Programme, with the following objectives (SMH, 1995):

- to bring the informal city as close as possible to the regular city, through integration and settlement upgrading;
- to transform *favelas* into neighbourhoods through upgrading, land titling and community and public services delivery;
- to expand the basis of urban property titles over *favelas*, residences, plots and housing estates; and
- to broaden building opportunities through the occupation of vacant land and available serviced land parcels.

Implementation was initially undertaken with municipal resources. However, the SMH later managed to mobilise finance from the Interamerican Development Bank (IADB) allowing the budget of the upgrading programme to be increased to US$ 300 million.

Selection Criteria for Upgrading

During the period of preparation of the *Favela Bairro* Programme, a methodology was developed to assist in the selection of the target settlements; selection criteria included:

- the size of the *favela*;
- the balance of constraints and possibilities for upgrading;
- the level of existing infrastructure; and
- the socio-economic needs of the inhabitants.

Due to the very high total costs that would be necessary to upgrade large settlements, 114 *favelas* with more than 2,500 families each were excluded. In addition, small settlements were also excluded due to high relative costs. The most cost effective size was determined to be between 500 and 2,500 households or between 2,000 and 11,000 inhabitants. After a survey using indicators of urbanisation was completed and the results analysed, 85 *favelas* were finally selected.

The Method and Process of Upgrading Favelas

The *Favela Bairro* Programme was intended to integrate *favelas* with the formal city, turning them into neighbourhoods and integral part of the urban structure. Physically, then, the programme has emphasised accessibility, the opening and paving of roads and defining public spaces. Most important, it has provided each settlement with an integrated urbanisation plan in which the settlement layout, including both public and private areas, is well defined. This is essential for the formalisation of land tenure and the citizenship of the inhabitants. The squatter residents thereby become part of the real estate property cadaster and subject to the city's taxation. At the same time, they receive the rights to demand from the municipality access to and the provision of infrastructure and public services. Citizenship implies rights and obligations for both the citizen and the State. Box 1 below summarises the expected results of the *Favela Bairro* Programme.

In 1994, the SMH in co-operation with the Brazilian Institute of Architects organised a public competition to develop upgrading methodologies. Planning and architecture offices were invited to present proposals and ideas to upgrade the *favelas*, and 34 different proposals were received. Fifteen of them were selected. The winners were then contracted to develop detailed upgrading plans in pre-selected settlements. These teams were responsible for approaching and working with the communities, for carrying out all necessary fieldwork, for developing concepts and initial plans, for formulating detailed design plans and for delineating action plans.

Box 1: Expected Results from the Favela Bairro Programme

- Integration of the *favela* with the city;
- Improved accessibility, roads, sanitation and public lighting;
- Organised solid waste collection and urban cleansing;
- Improved health conditions;
- Improved opportunities for community and social interaction through the provision of public spaces and community facilities;
- Support to children of pre-school age;
- Reduction in the risk of flooding;
- Improved internal and external environmental conditions; and
- Improved sense of citizenship and belonging to the city.

Source: SMH, 1995.

The SMH organised its staff in teams of project managers to supervise both the formulation of the plans and the actual implementation. The main role of these managers was to guarantee that the development of the projects and their implementation are carried out according to the established principles and norms for public tendering in Brazil. Project managers were the pivots between the planning teams (elaboration of projects and intervention plans), the building contractors who are awarded contracts to implement the projects (construction activities) and the community associations (final users and beneficiaries). The execution of the physical works has also been subject to public tendering; and the contractors were selected according to their experience, their capacity, their prices and their ability to manage and execute public works of this nature. The results of a typical project under the *Favela Bairro* Programme are summarised in Box 2.

Box 2: A Typical Favela Bairro Project

In October 1996, the municipality of Rio de Janeiro officially turned over the partially completed upgrading programme in the Serrinha *favela* to its inhabitants. A total of 5,000 m² of pavement (new roads, alleys and pathways), 2.4 km of sewerage network, 2.5 km of water supply network, 400 household connections to the sewerage system, 360 water points, 1.5 km of staircases, 1.8 km of drainage gutters, protection of slopes from landslides, as well as the resettlement of 2 families who were living in areas at risk, were implemented during the project. These interventions represented 70% of the entire project, with an investment US$ 4.365 million.

Source: SMH, 1995.

After the initial phase of implementing projects of this type and scale, other settlements were selected for an expansion of the programme. The great success of the *Favela Bairro* Programme played an important role in the results of the last municipal election. For the first time in history, the population re-elected the governing political party. The new mayor is an architect who once led IPLANRIO and who designed the city's rehabilitation programme (Rio Cidade) and promoted the *Favela Bairro* Programme.

The degree of satisfaction of the inhabitants of the *favelas* is high, and a significant part of the city's population approves of the urban transformations which were carried out by the previous local administration. Multiplier effects of the programme are already visible in the upgraded *favelas*, and they need to be monitored properly. Public investments are also generating private investments for improving buildings, and there is a notable response in terms of management and maintenance of the services and equipment provided. There are also many signs that the 'narco-traffic' is relocating to other sites. The opening of roads and better accessibility of the *favelas*, i.e., their integration with the surrounding neighbourhoods, result in more spatial transparency and easy access of the police and public security forces. An underlying, but not often explicitly articulated, objective of the *Favela Bairro* Programme has been to neutralise the influence of organised crime in the *favelas* of Rio and to bring back the presence of the State to these areas, reinstate governance and restore municipal authority, along with its laws, norms and values after decades of neglect. It appears to have been successful in doing so.

Concluding Remarks

The *Favela Bairro* Programme has focused on the general improvement of public spaces and collective needs from the perspective of environmental improvement. It has had the twin objectives of defining and improving leisure spaces, parks and areas for social interaction and sports, as well as of solving the problems of the brown agenda. The second is particularly important in that the majority of *favelas* dispose of their waste in natural drainage courses that lead to important water courses of the city. There are strategic linkages between the improvement of sewerage and drainage systems in the *favelas* and the pollution alleviation programmes of the main water courses and Guanabara Bay (financed by IADB and Japanese funds). Unfortunately, this integrated vision is not strongly present throughout the city government, nor among the citizens at large. This emphasises that the success of this programme is just one of the many that must occur if the management of a megacity like Rio de Janeiro is to achieve sustainable improvement.

Although there has been a high level of acceptance for the *Favela Bairro* Programme, it has followed the orthodox formula which assumes that public investments will generate private investments at the household level. In fact, in economic terms, the *Favela Bairro* can be characterised as a programme of public investment, and there is hardly any concern with cost recovery. The municipality has played the traditional role of provider, although it utilised the participation of the teams of architects and planners who have acted as consultants and intermediaries between the government and the inhabitants. Although there have been several public-private partnerships in the squatter upgrading programme, the municipality of Rio opted to recapture governance in an orthodox way. However, the lack of cost recovery threatens the possibility of extending the programme to most *favelas* and leaves unsolved a basic urban management problem.

The fact is that the population of Rio has recovered its optimism and responded positively to the revitalisation programmes, of which *Favela Bairro* is one. Crime has decreased, and there are more employment opportunities. However, the critical environmental problems of the city remain unresolved because they depend on the specification of roles and responsibilities of the city and other levels of government and good co-ordination among the 14 municipalities that form the metropolitan region. Solving urban environmental problems piecemeal within individual settlements provides no sustainable solution to the environmental problems of Rio, no matter how popular and successful any particular upgrading programme might be.

References

IPLANRIO (1993), *The City of Rio de Janeiro*, IPLANRIO, Rio de Janeiro.

JB (*Jornal do Brasil*) (1996), 'Prefeitos Colhem Frutos da Súbita Riqueza das Cidades', 18 August, Rio de Janeiro, pp. 15.

Mitlin, D. and Satterthwaite, D. (1994), 'Cities and Sustainable Development', background document for the workshops, presentations and discussions in *Global Forum '94*, 24-28 June, Manchester, United Kingdom.

SMH (Secretaria Municipal de Habitação) (1995), *Política Habitacional da Cidade do Rio de Janeiro*, Prefeitura da Cidade do Rio de Janeiro, Rio de Janeiro.

12 Solid Waste Management in Copenhagen

JEFF COOPER

Introduction

One of the problems in providing an overview of world trends in waste management in metropolitan areas is that often it is difficult to discern the main driving forces in a mass of statistics. Frequently the statistics themselves are as misleading as enlightening because they are based on different concepts in each country.

This chapter therefore will focus on the development of waste management systems in Copenhagen, the capital of Denmark, as representative of trends in many of the North West European countries. Denmark has a population of around 8 million people and a broadly based industrial sector with a strong agricultural component. Heavy industry is almost absent.

Each city, however, has its own individual perspectives. In many ways Copenhagen is at the extreme end of the spectrum because municipal resources are sufficient to deliver a very high standard of service. The population receives a very wide range of services and facilities, and workers have excellent working conditions.

In contrast, in the former Eastern bloc cities, where the pressure on resources of all types is severe, it is difficult to maintain services to the standard developed in the Communist era. In most of these Eastern bloc cities the amounts of waste which require disposal, on a per capita basis, are half that generated in Copenhagen. The correlation between higher per capita GDP and rising waste generation levels is hard to break. Even though Copenhagen has instituted a comprehensive range of recycling options the poverty of the former Communist counties has led both to less resources being used initially and an enormous incentive to re-use and recycle everything.

In every case it is necessary to balance the resources devoted to waste management compared to other municipal services. With waste

139

management the priority must always be the provision of a cleansing service which helps to maintain the health and safety of the city's inhabitants and their environment. Beyond that there are, as the example from Copenhagen shows, many useful services which can be provided, but these should be evaluated against other service demands placed on municipal budgets.

Waste Management Priorities

One of the issues facing all waste managers is to determine which types of waste should be prioritised and then the initial priorities for treatment. In the municipal waste sector managers are often constrained by the types of waste they are required to deal with. Notwithstanding, it is important to examine the whole range of wastes generated and the extent to which municipalities can control and manage that waste.

The relative amounts generated by the household, commercial and construction and demolition (C&D) waste sectors and their disposal outlets are shown in table 12.1. The table shows clearly the preponderance of C&D waste, with industrial and commercial waste exceeding the amount of household waste.

Table 12. 1 Types and methods of waste disposal in Copenhagen, 1996

Type	Landfill	Recycling	Incineration	Amount (Tonnes)
C&D	1	90	9	380,000
Commercial	-7	41	52	220,000
Household	3	19	78	180,000

Source: Cooper (1997).

The C&D waste is double the tonnage of household waste. The amount of commercial and industrial wastes is only about 20% greater than the household waste because in Copenhagen there is so little heavy industry, which in the typical Western city it is usually much more. These figures are for Copenhagen and the enclave of Frederiksberg only, not the Greater

Copenhagen area.

The main lesson from these statistics is that collected municipal waste is often much lower than that generated by industry and especially by the C&D sector. This pattern is common to all major metropolitan areas in developed economies.

While there is a requirement on municipal authorities to provide waste management services for household waste, the proper management of other waste streams also needs to be assured, both for environmental reasons in the case of C&D waste, and more for health and safety reasons in the case of commercial and industrial wastes.

Denmark has been successful in establishing a consensus as to how resources should be managed and waste should be handled. Thus the imposition of taxes in order to enhance the reclamation and recovery of waste is accepted by the population. This may be necessary to bridge the gap both with the raw materials production system which does not include external costs, as well as with the waste disposal system which also may not reflect externalities.

As a consequence because there has been an emphasis placed on reducing waste and recycling as much as possible, there has also been an acceptance of the need to have waste-to-energy plants. Even landfill is accepted because so much is done to avoid any landfilling. In the future, to avoid the landfilling of organic waste, only waste which is pre-treated will be allowed. Because of the high tax affecting landfill, even 60% of the ash from incineration plants is recycled.

The Administration of Waste in Copenhagen

Copenhagen's whole range of refuse collection and disposal services is provided by a not for profit company, Renholdingsselskabet af 1898 (R98). R98 is a co-operative, self-governing company. The Council of R98 consists of representatives of landlord's associations, tenants' associations, the Copenhagen and Frederiksberg Councils and R98 staff (Rasmussen, 1996).

Founded at the end of the Nineteenth Century to address the need to provide basic sanitation services from an expanding Copenhagen, R98 still collects night soil from premises in and around Copenhagen but has also established a comprehensive system of collection and disposal systems for solid waste. Increasingly the emphasis has been on collection for recycling.

The payment for refuse collection services is through fees which are collected together with other local taxes by the municipalities and then passed on to the company.

Many of the services provided by R98 are through joint ventures with other companies and some are through subsidiaries whose costs are ring fenced so that there is no subsidy from the municipal services. One venture is R98's one third share in the new bulk waste processing facility recently established to deal with Greater Copenhagen's waste.

The establishment of the Copenhagen Recycling Centre (Kalvebod Miljocenter) commenced in 1995. Located in the western part of Amager, south of Copenhagen it was officially opened on 2 May, 1996. The former more central but cramped C&D waste recycling site was vacated in 1995 to make way for preparatory works for the project which will link Denmark to Sweden by a tunnel and bridge.

The new site is 100 ha with three 7.5 ha special facilities for C&D waste recycling, green waste composting and contaminated soil treatment. The rest of the site will be used as a landfill, and though fully permitted, no waste will be accepted until the other landfill currently serving Copenhagen has closed down early in the new millennium.

The C&D waste recycling facility operated with a staff of 40 has a capacity of more than 800,000 tonnes and recycled 700,000 tonnes in 1996. Waste is accepted into the site at fees which represent the ease and costs of processing, ranging from Dkk 65 (£6) per tonne for pure concrete, Dkk 90 (£8) with ferrous reinforcement through to Dkk 410 (£39) for gypsum board and Dkk 710 (£68) for mixed C&D waste.

Equally there is a range of treatment processes to which the waste is subjected, ranging from direct delivery to the crusher for concrete waste, pre-crushing by a mobile jaw crusher for the steel reinforced concrete. The mixed waste is separated initially by mobile cranes with the resultant fractions going through the processing facilities which includes hand sorting of waste. Two thousand tonnes per year of steel reinforcement are extracted for despatch to a steel works in northern Denmark.

With a waste disposal tax of Dkk 285 (£27) per tonne when waste is landfilled there is plenty of opportunity to treat waste to avoid paying the tax. Also there is a modest Dkk 5 (£0.42p) per m^3 tax on sand, gravel and quarried stone used in construction which provides further encouragement to contractors to use recycled raw materials (MEM, 1997).

Table 12.2 Pricing of construction materials in Copenhagen, 1997 (Dkk per tonne)

Natural raw materials (gravel)	63
Tile (0-32 mm)	32
Concrete (o-32 mm)	52
Asphalt/concrete	49
Tile/concrete	24

Source: Cooper (1997).

The pricing policy for the recycled materials is affected by the price of competing natural materials (including tax) but overall the value is low, given the ready availability of both land based and marine dredged aggregates.

One of the advantages which is offered by the C&D recycling facility is substantial savings in movement of waste. In 1988, C&D waste was often transported 30-40 km to landfill, which were often the sites also used for the extraction of raw materials. It is estimated that this amounted to a total of 16.5 m tonne-kilometres while in 1992/93 this had fallen to 2.7 m.

The progress of recycling of construction waste in Copenhagen has been dramatic, with only a very minimal amount now going to landfill (Table 12.3). Equally, the proportion of construction materials accounted for by recycled materials has risen from 1% in 1988 to 19% in 1992 and 25% in 1996. It is likely to remain at that level for the future because so much construction materials go into the long term stock of buildings and structures.

Table 12.3 Disposal routes for C&D waste, 1988-1996 (percentages)

	1988	1992	1994	1996
Landfilled	84	10	3	1
Incinerated	--	8	10	9
Recycled	16	82	87	90

Source: Cooper (1997).

The composting plant is the largest in Denmark. In 1995 Denmark composted 535,000 tonnes of material, of which 434,000 tonnes were accounted for by green waste with household, sewage sludge and animal wastes accounting for approximately equal shares. The Copenhagen site has a capacity of 100,000 tonnes per year.

On arrival, green and parks waste is checked for contrary materials because there is a considerable price differential between pure green waste and contaminated material in order to deter those tempted to avoid tight segregation, a problem until the introduction of the differential prices at the end of 1996.

The waste once shredded is placed in windrows, turned once or twice a week with water added as necessary. The water comes from a 3 m deep tank filled by rainwater running off the surface of the site. The windrows are operated for 12-14 weeks and then the compost is sieved for size and placed in storage to mature. There is a range of compost products with the best quality, 0-20 mm, costing Dkk 80 per tonne. If there are too many wood chips for mulch, the excess is burned for energy recovery.

The soil treatment plant has a capacity of 45,000 tpa and handles predominantly oil polluted soils. The hydrocarbon contamination levels are typically 700-2000 mg/kg and have to be reduced to less than 50 mg/kg. If these levels are not reached, then the material can be consigned to a special landfill site.

The soils are mixed 50% by volume with freshly shredded green waste and placed into 1.5 m high windrows and turned once every week to ten days with water added as necessary. This means that the treatment can continue even during winter because a temperature of 15C can be maintained even with sub-zero air temperatures. While sandy soils are easily worked, clay soils are more difficult to clean, and in future it is planned to crush the clay so that air and water can be added. The product from the treatment, provided it reaches the required standard, can be used as a growing medium for embankments for newly constructed roads and other construction projects.

The Recycling Centre offers Copenhagen a superb facility for the treatment of bulk wastes which admirably complements its initiatives in recycling and energy recovery of household wastes.

Recycling

The collection of recyclable materials in Copenhagen is mainly through the provision of high density bank systems, mainly for paper and glass but even for organics from a limited number of apartment blocks.

Paper is collected mainly from apartment blocks and on street facilities. Plastics are not collected from household sources but, from 1998 in order to conform with the Packaging Directive, industrial and commercial holders of plastics waste will be required to ensure that it is collected for recycling.

Glass is collected in bottle banks but unlike other countries the bottles are collected and processed for reuse. This is only possible because of a substantial tax on new glass bottles manufactured in or imported into the country. As time has gone on, so the proportion of glass collected from bottle banks which is reused has declined and is now down to only one third of the total. Nevertheless, the bottle bank is essentially a fall back for a system which relies on people returning bottles to shops, usually with the inducement of a deposit refund.

Cans are not at present used for beverages in Denmark but this may change if the European Commission and/or Court rules this illegal. However, in an important case in 1986 the European Court ruled that the Danish "can ban" was justifiable on environmental grounds.

Metals are reclaimed from incinerators of which there are two serving the Greater Copenhagen area, with a combined capacity of 750,000 tpa, and from recycling centres, of which there are five in Copenhagen. These recycling centres are provided for the collection of a variety of bulky wastes, garden waste, hazardous wastes such as oil, batteries and fluorescent tubes. People bring in their wastes and segregate them into a number of containers for recycling, energy recovery or treatment.

Organic Waste Treatment

There are several types of organic wastes which can be treated to produce animal feed, growing media, and/or methane in the case in the case of anaerobic digestion. These are kitchen (bio-) waste, green waste from parks and gardens, and food waste from restaurants, canteens and other catering facilities. In Denmark, those facilities generating more than 100kg. week of food waste are required to arrange to have it collected for recycling into animal feed, mainly for pigs. Green waste is easily processed but there needs to be adequate moisture for decomposition and therefore water is often added. Compost from mixed refuse has been abandoned in northern Europe but has limited application in Southern Europe because of the limited amount of organic material available from other sources.

In Denmark, there was one plant for the processing of selected organic wastes, including paper, at Elsinore, but after three years of sporadic operation it was closed temporarily in June 1996.

Although home composting does occur in Denmark, it is not as actively promoted as in other countries where often free or subsidised composters are provided in order to divert waste from people's dustbins.

Innovation

There are several initiatives which the Danes have embarked on in order to ensure the health and safety of their workers engaged in waste management. For example, there is no hand sorting of household waste, even segregated dry recyclables, for example, because of potential detrimental health effects. Equally there is a concern to limit the workers' exposure to airborne micro-organisms which may be present in refuse, especially putrescible waste, such as source segregated bio-waste and kitchen waste.

There is a very rigid interpretation of the European legislation dealing with the lifting of heavy weights in the work environment which has lead to the Danes developing novel solutions to the problems of handling refuse.

For the past six years, R 98 has been installing suction systems for removing wastes from blocks of flats in various parts of Copenhagen but these have been serviced by a mobile refuse truck. The use of the mobile suction unit is increasing quite rapidly now after a very slow start. By mid 1997 there were 25 places with the facility and a further five planned. The system solves problems of access and handling refuse in older housing units with steps, low passage ways and cramped basement refuse storage areas. This is especially difficult issue in Denmark where regulations make it difficult for waste to be carried. Waste tends to be moved with the use of sack trolleys where wheeled bins are not used.

In modern housing developments the system has become increasingly popular because it occupies less space and causes less work for the landlord in the long run. The main barrier to its adoption is the high cost of installation.

The system is served by one mobile unit but this is due to be increased by a further unit because of the potential difficulties of breakdown and accident. The new unit will cost Dkk 2.2m (£205,000) more than twice the cost of a conventional refuse compaction unit. The new unit will have a 9m. suction arm to provide greater scope in the location of docking points.

The mobile suction unit will normally service each facility twice weekly with the whole process taking around 10 minutes from the time of arrival to departure. Everything can be controlled from the docking unit and vehicle, including the closure of the inlets of the tank and the sequence in which the individual tanks are opened and closed.

More recently a fixed facility was established in the Nyhavn (new harbour) area. It is a mixed commercial and residential area which has poor access for conventional refuse collection services. The pedestrianised area is thronged with people coming to use the bars and restaurant, especially in the summer (Cooper, 1997).

A 500 mm internal diameter steel pipe was laid under the road near to the harbour wall running 540 m through the area. There are 8 inlet points for the pipe and access to the inlets is available for 230 commercial companies and an equal number of residents living in flats.

Largely completed by May 1997, it was not fully operational until September 1997 due to reluctance of a small number of the restaurants to accept the new system. The total costs, including extensive development costs amounted to more than Dkk 10m (£1m).

At the specially constructed building housing the suction equipment and compacter, located near the ferry terminals, there was considerable expense incurred in ensuring excellent sound insulation and good design. The waste comes into the building, drops into a cyclone and is compacted into a 30m³ container. The container will hold up to 8 tonnes of waste and is exchanged whenever full, weekly in the depths of winter and occasionally - perhaps three times - in full summer.

These systems do, however, reduce both the opportunity and need to recycle waste and therefore while offering improvements in waste management it is at a cost of resource conservation. In the case of the fixed facility at Nyhavn this required special dispensation from the Government. Although recycling of glass and paper by commercial premises and residents is encouraged, no local facilities are provided. Wherever possible the places serviced by the mobile suction unit will also be provided with paper banks, while bottle banks are usually located at a greater distance from housing.

Oddly the Danes have, almost universally retained the paper sack for waste collection rather than go over to the plastic sack which has been adopted elsewhere, except where now superseded by the wheeled bin. The paper sack cost more than three times more than the plastic sack and presents a number of minor operational difficulties, especially in wet weather. Nevertheless, the use of the paper sack has been retained because it appears to be more environmentally friendly than the plastic sack. In Copenhagen special waterproofed small paper sacks have been developed for the collection of bio-waste from suburban households.

Conclusion

Solid waste management, as conducted in Copenhagen, is a complex activity involving a number of activities and a broad range of facilities. Solid waste management is given high priority by the municipal governments. It is well funded and a focus of urban management efforts. Both of these conditions reflect a high degree of consensus among the population that waste management is important.

It has become possible to achieve substantial recycling. Much of this has been the wastes of construction and demolition, which is the preponderant component of total wastes. Differential pricing and taxing is successfully used to encourage recycling. Composting is also carried out as well as recycling of household wastes.

Realities require compromise in some cases where recycling and collection are not entirely compatible. This has inspired innovative solutions requiring expensive technology which depend upon the substantial resources of a First World city.

References

Cooper, J. (1997), 'Denmark offers novel system in Copenhagen', *Materials Recycling Week*, vol. 169, no. 20, pp. 12-17, Emap Maclaren, Croydon.

MEM (1997), 'Affaldsafgiften 1987-1996', *Arbejdsrapport fra Miljostyrelsen*, no. 96, Miljo-og Energiministeriet Miljostyrelsen, Copenhagen.

Rasmussen, T.J. et al (eds) (1996), *Cleaner, Greener Supply Lines*, Copenhagen City Council, N Olaf Moller, Copenhagen.

13 The Social 'Nature' of Floods in Buenos Aires: Rainfall Increase or Higher Vulnerability ?

OSVALDO GIRARDIN AND GABRIELA GRECO

In the last years but mainly during the 1980s decade, several regions of Argentina were seriously affected by floods, which brought about grave social and economic consequences. From the point of view of some sectors of the scientific community, this situation mirrored at the regional and local levels the alleged global climate changes that have claimed international attention. In the case of Argentina's Wet Pampas region, the implied impacts include an increase in temperature and in the amount of rainfall (Hoffmann,1988; Hoffmann, 1975; Hoffman et al, 1987; Barrios et al, 1995; Girardin, 1996).

During that period, Buenos Aires experienced rainfalls of uncommon magnitude, such as those of 26 January 1985 (192.2 mm in 3 hours), 31 May 1995 (295.4 mm in 25 hours), and 16 December 1989 (82 mm in one hour) (Federovisky, 1990) which caused floods of catastrophic consequences. The immediate association of these floods with heavy rainfalls was used to argue that the floods were natural phenomena.

Indeed, several studies (Hoffmann, 1988; Hoffmann et al, 1987; Hoffmann et al, 1997; Hoffmann, 1975; Barros et al, 1995) show that, during the last decades, a noticeable increase in rainfall has taken place in several regions of the country, including the Metropolitan Area of Buenos Aires. They argue that the increase signals a kind of climatic change and not just as an climatic anomaly, since it occurred over a substantial time period. Information provided by the National Meteorological Service is shown in Table 13.1.

Table 13.1 Rainfall in Greater Buenos Aires

Decade	Annual average rainfall (mm)	Annual rainfall in 1981/1990 compared to each decade (%)
1911 - 1920	1,101	+ 9.7
1921 - 1930	986	+ 22.5
1931 - 1940	1,016	+ 18.9
1941 - 1950	975	+ 23.9
1951 - 1960	1,089	+ 10.9
1961 - 1970	1,076	+ 12.3
1971 - 1980	1,143	+ 5.7
1981 - 1990	1,208	---

Source: Authors' calculations with data supplied by the National Meteorological Service.

According to this data, the decade of 1981/90 was, on average, the wettest up to now in the century, and there was 21.7% more rain than the average of the period from 1921 to 1950. However, it is evident that the average of this last decade was only 5.7% more than that of the period 1971/80 and only 9.7% more than the mean of 1911/20. This last point, together with Federovisky's (1990) report that there were floods from rainy spells that did not surpass usual values, challenge the belief that heavier rains adequately explain the recent floods in Buenos Aires City.

The Growth of Buenos Aires

Many cities of Latin America have large areas which become flooded. Strangely enough, the oldest sector of those cities, that one which was founded by the Spaniards, does not get flooded. The customs of the indigenous people required that settlements should be protected from flood dangers. In the case of Buenos Aires, its settlement was limited to the highest part of the plain.

The evolution of the configuration of human settlements depends on the models for economic development adopted by each country (Vapnarsky and Gorojovsky, 1990; Vapnarsky, 1995; Soler and Rubio, 1992; Hardoy, 1993; and Girardin, 1995b). In the Argentine case, the process of incorporating the Wet Pampas lands into the world livestock trade, following a model of agricultural exportation, produced a strong concentration of human settlements in the Littoral areas, especially those of Buenos Aires and its surroundings. Such economic forces in the context of the country's geomorphology, created the kind of society and urban system which is characteristic to Argentina (Vapnarsky and Gorojovsky, 1990).

During the first years after the foundation of Buenos Aires, land rights were allocated unequally regarding plot size and location qualities, establishing a pattern which continued to the present time. The use of slaves for manual labour, the development of various smuggling routes, and the consolidation and strengthening of a trading class, led to a diversification of the social structure which is reflected in the use of urban space. Later on during the second half of the 19th century, and from the abolition of slavery, the process of land occupation accelerated, mainly on low-lying, floodable areas. This was done by the freed men whose change of the legal status allowed them to relocate, but to do so outside the central area of the city. They chose to occupy the floodable areas of the Lowlands (El Bajo) as well as the Terceros basins. In contrast, the urban elite established themselves in flood-free areas.

This combination of the city's expansion with an increase of social differentiation brought together the natural risk of floods and particular social groups (Lindon, 1989).

Some Consequences of the Process of Urbanisation

Urban agglomerations constitute one of the most drastic ways in which humanity alters nature. Urbanisation processes introduce great modifications to the natural dynamics of hydrological cycles, affecting the three key parameters of the drainage, infiltration, and interception coefficients. These changes arise from high concentrations of built-up areas, and are worsened by the city's size and the lack of such planning as would mitigate the effects caused by population concentrations.

Di Pace et al (1992a and 1992b), Hardoy (1993) and Vapñarsky (1995), mention some of the aspects which suffered negative impacts (still present now) because of policies applied to the structural adjustments of Argentina's economy. Those policies impacted both the poverty sector as the urbanisation processes. A tendency is seen actually towards a growing spatial marginalisation of the poorest sectors of the population, which is manifested by their continuous displacing to the most environmentally vulnerable areas.

Many interrelations are being established in the course of this fusing process between the impacts derived from poverty and environmental degradation, mostly amongst the different urbanisation processes and development levels, income distribution's heteregoneity and environmental impacts.

It is not the objective of this document to make a detailed review of the process of the growth of Buenos Aires Metropolitan Area. Nevertheless, it should be noted that all studies agree that the main characteristics of this process were chaos and lack of planning (Del Giudice, 1994; Brailovsky and Foguelman, 1992; Hardoy, 1993). Hardoy (1993) outlines how Argentina's urbanisation occurred without the the required investments in the infrastructure for supplying essential services (drinking water, waste water removal, etc.). Neither was any consideration given to the environmental matters, such as the way hydrological cycles were affected by surface drainage or a decrease in the quality of drinking water. Modifications to the land and the environment of suburban areas altered the natural flow of water, the demand for land and building materials used. Basic studies were not done which would determine if the land was suitable for the establishment of human settlements.

The lack of planning resulted in a huge city with few open and green spaces and with a low density, which increased the cost of laying out the different public services networks for providing drinking water, sewerage, electricity, etc. Access to service infrastructure depends on two key factors: nearness to the Capital City and income levels. At the same time, low incomes make some social groups more vulnerable to threatening events. This is because their poverty reduces their capacities to absorb, alleviate or adapt to the consequences of a catastrophe.

Urban Growth and Floods: The Case of Maldonado Stream

Latin America's urbanisation has its greatest expression in three metropoles, namely Mexico City, Sao Paulo and Buenos Aires, whose explosive growth made vast areas impermeable to the natural infiltration into the ground of rainwater. At the same time, all this process has been accompanied by an almost total loss of the vegetal cover which eases the infiltration process, by withholding rain water and by the extraction of great amounts of soil for building processes which created large pits. It must not be forgotten, that, in spite of the artificiality of an urban area, ecological processes are not at all annulled, but rather they continue and are manifested in different ways.

In Buenos Aires City all water courses originating in the rest of the Metropolitan Area are collected together. They carry all rainfall water which cannot infiltrate into the ground and discharge this into the Rio de la Plata and the Riachuelo.

Although the city does not have problems of instability caused by slopes, as is the case for many Latin American cities, it does have difficulties as regards its drainage system, as a consequence of the mild slopes of the Pampas Plain on which it is situated. Therefore there are natural flood areas, such as those along the rivers and streams crossing the city, as well as the banks of Rio de la Plata. Furthermore, this river is naturally influenced by tides which, when high, dam the river up, which is therefore not able to discharge its water into the sea for some hours. A similar effect is created by the rainy southeast wind blowing up the river which slows down the rate of drainage (Brailosvsky and Foguelman, 1992).

Excessive urban growth, which was not backed by any planning process, has revealed the vulnerability of some of the areas with this characteristic, as described in the following:

> ... the lower valley of the Maldonado stream was considered as a highly risky, but not a vulnerable region as regards natural floods during the 16th and 17th centuries; however, from the end of the last century, and because of human settlements, its vulnerability became clear, while the risk of flooding was always the same, before and after land occupation. This means that the risks can be attenuated or aggravated, according to the conditions under which the occupation of the land takes place (Lindon, 1989).

Federovisky (1990) argues that the construction of underground channels are now the main cause of floodings in Buenos Aires City. The Maldonado basin, which crosses the city almost through its middle, discharging its waters into the Rio de la Plata, is the biggest of the basins which is drained by underground channels. Half of it is located in the outskirts of Buenos Aires City in La Matanza district. It was developed as part of a drainage project for Radio Nuevo in 1919, extending from Rio de la Plata to the limits of Buenos Aires City.

The vulnerability of the Maldonado basin in Buenos Aires City was demonstrated when the underground channels were extended to the outskirts of the city, with a corresponding increase in drainage coefficients, because of the almost total impermeability of the watershed which this created. Thereafter, the volume of water more than doubled after an average rainfall. Some years later an alleviating channel was built in the direction of the Cildañes stream, one of the Riachuelo's tributary, which diverts only a minimum part of the overflow in the case of an average rainfall. The construction of two additional channels on the lower course of the stream was projected in 1935, but never carried out.

> ... the present drainage capacity of the Maldonado (338 m^3/sec) is amply surpassed by the average amount of rains used when calculating the sizes of the original piping system, given the increases on the draining coefficient. (Federovisky, 1990)

The situation became more complicated because of the accumulation of rubble and different types of rubbish in the channels, particularly in recent years. There was an invasion of a new kind of garbage produced by followers of the new consumption patterns, based on the use of discardable products, such as packages of all kinds, bottles, cans, and nappies. The presence of accumulated debris and rubbish may have different impacts on floods depending on the basin involved. Although this debris and rubbish obstruct the channelled water courses and sewers, increasing the overflows, they work also as retention dams of a kind in the higher water basin, thus allowing the drainage of the lower basin.

Consequently, efforts to clean the sewers and remove debris and rubbish accumulated in the channels are not the solution of the total problem, even though it alleviates the situation of some areas which used to get flooded. When these obstacles are removed without resolving the structural deficiencies of the drainage systems for the waterbasins, the drainage velocity increases, causing water to accumulate in the lower basin. Here the slope gradient is small and there is less drainage capacity, so water becomes stagnant in small hollows within the valley, with the consequences that regions which before had no problem as regards flooding, begin having them (Federovisky, 1990).

Institutional and Legal Aspects

Flooding in Buenos Aires is affected by the lack of general environmental policies for Argentina which would provide and all encompassing framework as well as administrative tools for the elaboration and implementation of any policies and plans for the Metropolitan Area. This deficiency complicates and hinders the coordination between the different governing levels (National, Provincial and Municipal), giving place to overlapping as regards responsibilities and jurisdictions.

This situation is complemented by a trend toward dealing with the subject in a sectoral and fragmented way. This fragmentation involves both temporal and social aspects. On the one hand, action at the time of the catastrophe is separated from action before the next catastrophe (including prevention activities). On the other, social aspects are taken into consideration only during the moment of the catastrophe and only in terms of the social impact of the disaster. Their role in creating the disaster is ignored. There are no common guidelines for dealing with this fragmentation among the various jurisdictions, nor are there institutional mechanisms for legal and administrative coordination. There are no devices for ensuring adequate participation by the communities in the design and the operation and control of policies (Del Giudice, 1994).

When dealing with water resources, the fragmentation and overlap of responsibilities among various jurisdictions are sources of particular difficulty because of the lack of a common understanding of a water basin's system and an effective means of coordination. For example, in the case of the Matanza-Riachuelo basin, there are at least 34 different jurisdictions involved: the Nation, Buenos Aires Province, the City of Buenos Aires, several municipalities, several government agencies, such as the Secretariat of Natural Resources and Sustainable Development and the Secretariat of Public Works.

In Argentina the municipalities are given the means to formally create regulations, undertake works, and supply services. In this context, a municipality can be seen as the political and administrative organisation which manages local urban matters (Pirez, 1991; Pirez, 1995).

For their part, urban services, basic infrastructure and all kind of service networks constitute essential collective capital for the city's operation. Because they deliver public goods which sometimes are indivisible, it is not always easy to identify costs and benefits to individuals. Additionally development of these networks can offer strong economies of scale.

These characteristics call for an entity to organise coherently all activities of the metropolitan area, establishing the distribution of the service costs, settling the conflicts and clarifying responsibilities among various jurisdictions, and judging the heterogeneities between the different social groups and Municipal boundaries (Coing, 1989).

The total of all these deficiencies, together with the lack of comprehensive foresight make it very difficult to counter the dangerous consequences of flooding. Since Buenos Aires is located downstream in the basin, it suffers the consequences of upstream activities. The area functions as a whole, as regards climatic, hydrological and social aspects. It is independent of political-administrative boundaries and jurisdictional limits. Therefore, the region's overall functions have to be taken into consideration when trying to fight the problem.

Conclusions

More than natural factors associated with floods in Buenos Aires' Metropolitan Area, it seems that disasters are the result of economic, political, and social conditions.

The 1981/90 period has been the wettest one of the century, but the magnitude of rainfall increases (5.7% compared with 1971/80 average) cannot by itself explain the catastrophic effects of the floods which occurred on 31 May, 1985 and 16 December, 1989.

To discover the real reasons that transform floods into disasters, one should be look into the explosive and unplanned urbanisation process without the support of essential additions to the infrastructure system. Between 1914 and 1991, the total population of the Buenos Aires Metropolitan Area increased from 3.7 millions to 14.3 millions, an increase of 285%. The corresponding increase in Buenos Aires City from 1.6

million to nearly 3 million was nearly 90%, while the remainder of the Metropolitan Area increased during the same period by 433%, from 2.1 millions to 11.3 millions. Between 1947 and 1991, the population of the total Metropolitan Area increased by 84%, while that of Buenos Aires City decreased by 0.7% and the rest of the Metropolitan Area increased 137% (Hardoy, 1993).

During that growth, not only were the natural dynamics of hydrological cycle modified in terms of three of the main parameters (drainage, infiltration and interception coefficients) but also a continuous process of spatial segregation of social groups took place as urban space was appropriated, thus joining social differentiation with the natural risk of flooding to create unusually negative consequences.

As stated by Lindon (1989), the flooding problem is not only a consequence of the occupation of vulnerable areas by the poorest social sectors (although areas occupied by medium and high income sectors of the population are also affected by this problem), but the incidence as well as the ability to stand up to the effects of these events are, indeed, quit different. In other words, the vulnerability of the different social sectors is very dissimilar.

In fact, floods point out the high vulnerability levels of extensive groups of the population, showing up their poverty and their lack of alternatives. At the same time, floods reveal the obsolete state of most of the existing urban infrastructure, thus making clear the absence of proper actions to address pre-existing situations (Herzer, 1990).

However, the necessary solutions are difficult to implement. In the first place, there is no possibility to clear away those already living in flooding risk areas (Lindon, 1989), due to the conflict between needs for urban space (meaning space that can be inhabited without any risk and where the people would be able to develop their common activities without interruptions), and the actual availability of such urban space. Moreover, the lack of an agreed approach to a water basin and an institution able to coordinate actions makes it very difficult to find a solution to the problems. Treatment of problems is further complicated because flood plain boundaries and political-administrative ones do not coincide.

Probably, a necessary next step to be followed up is to establish an agency able to deal with the whole area, one which is able to resolve the fragmentation of competence and responsibility among the diverse national, municipal and provincial bodies. In addition to adequate resources, such an agency must have, of course, the ability to define priorities, practices and timetables for action as well.

The problem of the quite often severe consequences of flooding in Buenos Aires City leads us to be aware that society, in some circumstances, loses control, completely sometimes, of its own living space. Furthermore, if this living space is considered the result of historical and social processes embedded in its construction, it must be seen as more and more complex, more segmented and shaped by an extensive variety of actors, and consequently less understood and less manageable by mankind (Greco, 1995).

References

Barros, V., Castañeda, M. and Doyle, M. (1995), 'Recent precipitation trends in Southern South America to the East of the Andes: An indication of a mode of climatic variability', in Rosa, L. P. and Dos Santos, M. A. (eds), *Green House Gas Emissions Under Developing Countries Point of View*, COPPE/UFRJ, ENERGE, ALAPE, Rio de Janeiro.

Brailovsky, A. and Foguelman, D. (1992), *Agua y Medio Ambiente en Buenos Aires*, Editorial Fraterna, Buenos Aires.

Coing, H. (1989), 'Privatización de los servicios públicos: Un debate ambiguo', in Schteingart, M. (ed), *Las Ciudades Lationamericanas en la Crisis. Problemas y Desafíos*, Trillas, México.

Del Giudice, F. J. (1994), *Guía Ambiental de la Argentina*, Espacio Editorial, Buenos Aires.

Federovisky, S. (1990), 'Influencias de la Urbanización en un Desastre: El caso del Area Metropolitana de Buenos Aires', *Medio Ambiente y Urbanización*, no. 30.

Girardin, L. O. (1995), 'La reestructuración de la economía mundial, la desconcentración industrial y sus eventuales efectos sobre el sistema de asentamiento urbano en la Argentina. Las aglomeraciones de tamaño intermedio y su posible evolución futura', Fundación Bariloche. Programa de Medio Ambiente, Buenos Aires.

Girardin, L. O. (1996), 'Los gases de efecto invernadero y los cambios climáticos globales. Los riesgos de la toma de decisiones en un contexto de incertidumbre', Fundación Bariloche, Programa de Medio Ambiente, Buenos Aires.

Greco, M. G. (1995) 'Iruya: Un largo camino de trashumantes', in *Investigación, Conservación y Desarrollo en Selvas Subtropicales de Montaña*, Brown, A. and Grau, H. (eds), LIEY, UTN.

Hardoy, J. (1993), 'Urbanización, sociedad y medio ambiente', in Goin, F. and Goñi, R. (eds), *Elementos de Política Ambiental*, Honorable Cámara de Diputados de la Provincia de Buenos Aires, La Plata.

Herzer, H. (1990), 'Los Desastres no son tan naturales como parecen', *Medio Ambiente y Urbanización*, no. 30.

Hoffmann, J. A. J. (1975), *Atlas Climático de América del Sur*, MWO/UNESCO, Budapest.

Hoffman, J. A. J., Nuñez, S. and Gomez, A. (1987), 'Fluctuaciones de la precipitación en la

Argentina en lo que va del siglo', II Congreso Interamericano de Meteorología, V Congreso Argentino de Meteorología, Buenos Aires.

Hoffmann, J. A. J. (1988), 'Las variaciones climáticas ocurridas en la Argentina desde fines del siglo pasado hasta el presente', in *El Deterioro del Ambiente en la Argentina*, FECIC, Buenos Aires.

Hoffmann, J. A. J., Nuñez, S. and Vargas, W. (1997), 'Temperature, humidity and precipitations in Argentina and the adjacent sub-Antartic region during the present century', *Meteorologische Zeitschrift*, Revista científica de las Sociedades Meteorológicas de Alemania, Austria y Suiza.

Lindon, A. (1989), 'La problemática de las inundaciones en áreas urbanas como proceso de ocupación. Un enfoque espacio-temporal. El caso de la ciudad de Buenos Aires', Actas del II Encuentros de Geógrafos de América Latina.

Pirez, P. (1991), *Municipio, Necesidades Sociales y Política Local*, Centro Editor de América Latina / IIED-AL, Buenos Aires.

Pirez, P. (1995), 'Actores sociales y gestión de la ciudad', in *Ciudades*, Red Nacional de Investigación Urbana, no. 28, México.

Soler, F. And Rubio, G. (1992), 'Efectos espaciales de la actividad frutihortícola de exportación', *Revista EURE*, vol. XVIII, no. 54, pp. 65-78.

Vapñarsky, C. A. and Gorojovsky, N. (1990*), El Crecimiento Urbano en la Argentina*, IIED-AL, Centro Editor de América Latina, Buenos Aires.

Vapñarsky, C. A. (1995), 'Primacía y macrocefalia en la Argentina: La transformación del sistema de asentamiento humano desde 1950', *Desarrollo Económico*, vol. 35, no. 138.

14 Solid Waste Management in Colombo

V. U. RATNAYAKE

The problem of poorly managed waste collection and disposal has now reached every doorstep in the densely populated Colombo metropolitan area. The matter has reached such national dimensions that it occupied the top most position on the political agenda during the last local government elections in Sri Lanka.

What is Waste?

The composition of the solid waste in Sri Lanka is not similar to that seen in developed countries due to socio-economic and cultural reasons. Municipal solid waste is mainly garbage comprising household refuse and discards from smaller trade premises. Within the country the nature of the household refuse varies considerably depending on the income level of the occupants. Only the waste of the more affluent residents of Colombo approaches the composition of average European household refuse. The major component of the Colombo household garbage is organic matter of which about 99% is naturally derived perishables. However some percentages of glass, ceramics, food wrappings, fresh plant parts, and used garments are also found in the municipal garbage, perhaps a major portion of which is arising from trade premises. It has been estimated that only 85% of the garbage is combustible as well as suitable for land filling.

Hospital waste is one other component of the garbage collected by the Municipality as a consequence of bad material management procedures for the disposal of clinical wastes: dressings, sharps, swabs and outdated pharmaceuticals together with the associated packing material. The entire collection of hospital waste is transported by the Colombo City Council following the same procedures using the same garbage trucks and routes as for the general municipal waste. There are occasional complaints about animal carcasses, fish and meat offal being dumped in the open within the council managed cemeteries. Their decay causes a major public nuisance arising from nauseating smells. In certain instances these deposits are set

163

on fire, creating further havoc due to evolution of obnoxious smoke.

The waste generated in Colombo has not been measured. The per capita generation figure has been reported as 0.98 kg/day by some freelance social workers but without any supporting scientific data. This figure has been calculated on the basis of the permanent resident population, but it presumably includes waste generated by Colombo's commuting population which swells the number of people in the city daily by about 45%. According to the National Transport Commission's unpublished estimates, at least 2 million workers and school children come to the city daily and almost all of them generate some garbage out of the personal belongings they carry with them. The per capita waste generation figure also includes for commercial, industrial and institutional waste. The author estimates the a more realistic waste generation figure for Colombo's 4.5 million inhabitants (including infants) would be around 0.5 kg/day/person, which is calculated from the number of garbage truck trips made and a total of 850 mt waste generated, as reported in a private interview with the Council engineer in charge of the collection fleet. The author also estimates that the annual growth rate for per capita waste generation within the city area could be taken as 1% until the next Millennium as the urban Sri Lankan society becomes more consumer oriented.

Many items of no commercial value are also transported to the city daily by commuters and the trade. A very good example is the case of bananas. They are carried into the city while still on their long curved stalks, each of which may weigh around 10 to 15 kilograms and has no significant commercial or food value. Thousands of such stalks come to the city daily. At the end of the day, it is the City Council which is burdened with the task of transporting them away to their dumping sites.

The story with regard to some vegetables is similar. Grown on outstation farms, they are transported to the city markets daily with minimum facilities for proper packaging, transport and storage. According to records available, a considerable amount of vegetables is destroyed in transit and this figure runs into thousands of tons annually. This particular portion of the market refuse alone constitutes a considerable percentage of the total daily refuse of the city.

Linked Environmental Effects

The lack of dustbins at suitable places for public use results in most refuse discarded on pavements, and in corridors and alleyways by daily commuters and pavement hawkers. Eventually this finds its way into roadside drains, ditches and ducts causing a series of problems to the authorities. During cyclonic winds and torrential rains the resulting clogging of municipal drainage creates flooding. The problem is enhanced by the lethargic attitude of garbage collectors employed by the City Council. Recently the situation has been made worse by the organised stealing of road manhole covers for sale as scrap cast iron.

The blocked drains result in the frequent flash floods, even during short spells of torrential rains. They also create stagnant water pools along the drains which become breeding grounds for mosquitoes. In turn, the Council is compelled to spray the drains and the air with strong pesticides to the tune of 400 kg per month in order to reduce mosquito population and incidences of vector borne diseases.

A recent outbreak of a mosquito-borne dengue fever shocked the inhabitants of the entire city by causing a number of deaths of children under 13 years of age. A campaign was launched vigorously to control mosquitoes. Increased doses of insecticides were sprayed during the outbreak and a large number of blocked drains were cleaned. This eased the situation temporarily, but the toxic chemical agents used are reported to produce long term adverse effects on fish, other aquatic creatures living in water, birds flying about in air and the soil bacteria which maintain an ecological equilibrium. During this period the pressure on the health care delivery system was also increased to the extent that health authorities were taken by surprise at the increased demand.

Waste Collection

From the colonial days of Sri Lanka, local authorities have been charged with the responsibility of sweeping the streets, collecting and disposing waste by Municipal Ordinance No. 16 of 1947. Although the broader function of the local authorities is to look after the well-being of the citizens and to maintain a cleaner environment, the job of garbage disposal has not been performed as a part of an overall environmental management plan. The Municipal Council of Colombo experiences great difficulty in fulfilling the task of garbage disposal due to lack of trained garbage collectors, committed managers, adequate institutional infrastructure, incinerating or sanitary disposal sites, full public participation and political will.

The Municipality possesses around 165 rubbish trucks, each with a 6 ton holding capacity, and more are on order with built-in mechanical loading and compacting facilities which allow full utilisation of the vehicle's carrying capacity. Occasionally, open tractors and trailers and buggy-type lorries are also used for this purpose during any service breakdown or when the trucks are in concentrated use elsewhere. The use of tractors is popular on narrow roadways where the 6 ton garbage trucks are likely to create traffic hold ups. However none of these alternative vehicles has as much as a third of the capacity of a truck.

In areas where vehicular access is difficult, household waste is collected by handcarts operated by the Council employees. It is then deposited temporarily for about 5 to 6 hours at more than 100 convenient collection/transfer centres, until the garbage trucks collect it. In more developed areas, the Council has installed a large number of containers at selected sites so that people and the small trade premises can deposit their waste in them. These wheeled receptacles, having a capacity of 1 m^3, are made such that their contents can be unloaded into the garbage trucks mechanically. However, these are without lids and the contents are open to vermin, pests and birds. Birds pull the contents out of the containers. Other factors create an even more unsanitary mess at these sites. Habitually many people attempt to throw garbage bags out of motor car windows into these containers, but most of the time they miss, so that bags of garbage lie about, usually ruptured and spilling their contents. Garbage spills are not removed from the sites by the Municipal employees when the containers are unloaded mechanically.

Instead of these containers, three sided immovable concrete bunkers have been installed in poorer areas of the city to receive deposits of wastes from the collectors as well as from the householders. The garbage from these stations is loaded into trucks manually using spades. The surrounding areas are comparatively more untidy mainly because of the failure to clean the leftovers after unloading the bunkers into the trucks. Council workers have not been equipped with the kind of tools required to efficiently clean these stations either.

In certain areas within the city, waste collection centres as well as small scale waste disposal centres are on open grounds. Kept unattended, these attract the usual army of stray animals, pests, vermin and birds, and the neighbours traditionally set fire to partly dried garbage heaps in order to control these infestations and to avoid any foul smells likely to emanate from fermenting substances. However, this act unavoidably creates plumes of smoke for long hours of the day.

Sometimes the council has also distributed bags and buckets which the public can fill with waste and hand over to the collectors, but this facility has now been withdrawn due to protests from environmentalists claiming that the council promotes more polythene usage.

Disposal of Garbage

The garbage in the collection centres is taken by trucks to various sites where it is sorted out for possible reuse, sale or recycling by scavengers moving around the trucks as they unload and spread the garbage. Although such scavenging is not allowed officially, the authorities do not object to because the removal of unperishable material from the main garbage stock helps the garbage to decompose fast.

Some disposal sites receiving Colombo's refuse are unfortunately in low lying areas and only in some small sites has satisfactory action been taken to cover, compress and consolidate the refuse with soil to prevent large scale filtration into the underground water table of harmful leachate possibly containing toxic chemical agents in concentrated form.

Identification of waste disposal sites within the Colombo Metropolitan area with good access to transport vehicles has not been an easy task due to two main factors. First, the greater Colombo area is densely populated and no more space for extra infrastructure facilities is available. Second, residents already settled in the housing areas collectively and vehemently oppose setting up waste disposal sites in their vicinity because of the nuisance it is likely to create thereafter during waste disposal operations.

The main dump site used by the Council to dispose of the larger portion of the garbage collection has not been prepared for waste disposal at all. There are no installed facilities for leachate or gas release control; there are not even monitoring systems. Although peripheral ditches have been constructed to collect contaminated surface water at certain points within the site, there is no guarantee that the surface water does not penetrate into the local aquifer. The operation of the vehicles during the rainy seasons is also difficult due to lack of hard road surfaces. During rains the only alternative left is for the vehicles to tip their consignments near the entrance to the dumping site.

In such sites no cover material is spread over the waste in order to prevent the pests, birds and stray animals from encroaching on the site. The site releases considerable bad odour which has stimulated people living in the area to make frequent unheeded complaints to authorities against the operations. A number of residential facilities and a school are situated in close proximity to the site. There are reports that a chemical agent is sprayed to control the breeding of flies within the disposal site to prevent them reaching the local residential areas. During the rainy season a substantial quantity of deadly leachate is likely to percolate down through the dump into the ground water. According to the Council workers, soil for covering waste dumping grounds cannot any longer be sourced from nearby places as more and more areas are used by real estate developers for housing schemes.

Since the total volume of hazardous wastes generated by the Colombo group of hospitals is relatively small there was a proposal with sufficient funds to build an incinerator to burn hospital waste. This project did not materialise mostly due to the negative reasons mentioned above. Also the hospital could not identify a proper location to set up an incinerator.

Constraints to Proper Waste Management

Probably the most important constraints to proper management of solid wastes in Colombo are the lack of public awareness and the commitment to keep the city and its suburban satellite townships clean and tidy. These failings are not confined only to the city dwellers, but are applicable equally to policy makers, city fathers, politicians, public servants and daily travellers to the city. Policy makers have probably been viewing the current situation from their infant days and have come to accept it as a given. Additionally, no one has been assigned to be responsible for changing this poor state of affairs in the capital. A campaign of raising awareness among the garbage collectors has been resisted by the city administrators because of their shared belief that the workers are too poorly educated to be changed by it.

The lack of officers the responsibility of maintaining cleanliness of their own areas and answerable to rate payers of the Council has hampered the effectiveness of garbage collection.

Surprisingly, there exist no incineration facilities for treating solid municipal garbage in Colombo or anywhere else in the country. The main reason for the lack of interest to set up incineration facilities is the view held by some that the solid waste collected is too wet to be burnt at a low cost. The counter argument that it can be burnt after pre-treatment with solar drying finds no proper response. There is yet another group which claims that the incinerator flue gas contains poisonous furanes likely to harm public health.

Better Options for Improved Waste Disposal

Cleaning up at the end of the production cycle is fast becoming redundant approach as the concept of preventive waste management and control gains favour. This includes such things as recycle, reuse, recovery and, more recently reduction of waste at source. These ideas are now being recognised as a viable approach by the government to solve the garbage issue.

Reducing waste generation at site will certainly involve controlling the transport of all unnecessary material to the city markets by traders, encouraging daily commuters to the city not to carry non degradable food wrappers and throw them away in the city, minimising the wastage of both prepared and unprepared food at home and in commercial outlets, and separating out all reusable and recyclable material like textile off-cuts, glass, metals, and plastics. Waste minimisation does not require additional funds other than those needed to create greater awareness among all concerned of the importance of the waste minimisation practices.

Although the recycling of glass ware, polythene shopping bags and other plastic containers is carried out, the substitution of recycled materials for raw material in manufacturing processes does not appear feasible at this time because the recovery costs exceed the price of virgin raw material. An increase in import duties for raw material which competes with the recycled plastics and glass could change this situation.

At the same time, public awareness programmes should be conducted to convince the general public to exercise greater responsibility and discipline, to help Municipal workers to realise their responsibilities, and to encourage administrators to responsibly manage their collective efforts.

Improvement of the management skills of those responsible within the city administration is also needed. They should also be made more accountable to the Council but given more responsibility and authority. They need to adequately empowered and trained to take waste offenders before courts of law.

Although there are no sanitary land fill areas existing in Colombo or anywhere else in Sri Lanka, one site suitable has been identified approximately 30 km away from the main city centre. Authorities held a series of discussions, arbitrations and public education sessions in order to win the hearts and minds of the residents living near to the proposed site in order to get their consensus. If used, the number of garbage trucks has to be increased substantially because more driving time is required to reach the site.

Incineration of garbage is one of the best options for a country like Sri

Lanka where the available land space to set up landfill sites are rare, specially in greater Colombo area. Although the incinerators are expensive to set up, they are not as costly as land fill sites and are easy to operate and maintain.

Conclusion

Solid waste management in Colombo is being done in a haphazard manner that has become too complicated for the municipal council to manage. Given the steady increase in the resident population and ever increasing number of daily workers and school children commuting to Colombo, the Council's waste disposal operations may come to a grinding halt if its approach and techniques are not reviewed and improved.

The story starts with the transportation vegetables and fruits along with traditional packing material, indigenous types of containers, wooden boxes etc. from far-out farmlands to the down town markets. A considerable amount of this material gets damaged during the journey and becomes unsuitable for sales resulting in losses of valuable national assets running into few tons a day. Additionally a further load of waste is created by daily commuters to the city, from trade premises, office buildings, hospitals, garages, eating houses etc. Since the local authorities in the country are traditionally involved in street cleaning, the general public believes that it is the duty of Colombo Municipality to remove and dispose of garbage unilaterally and in isolation. However, the Municipality does not have adequate staff, a proper place to dump the waste, or any incineration facilities.

In the absence of an efficient system of waste management, uncollected solid wastes block the drainage systems and natural water paths. During even short spells of torrential rains, the stagnant water pools are created where mosquitoes breed and spread diseases. When outbreaks occur, the Council is compelled to spray large quantities of insecticides to control the mosquitoes, creating further problems of environmental pollution.

These circumstances call for both the local and central government administrators and policy makers to cooperate in tackling the problem as a national issue. This could make it easier to introduce new approaches and strategies as a part of the management of the total urban environment, rather than looking narrowly at solid waste management in isolation. Above all, a programme of public education is needed so that the local authority is no longer expected to carry out solid waste recycling, collection and disposal on its own, but can enjoy the support of a general public that wants to take art and knows how to help.

15 The Working Group Approach to Environmental Management under the Accra Sustainable Programme

BEN K. DOE AND DORIS TETTEH

Introduction

The involvement of all kinds of stakeholders through the working group approach is a primary feature of environmental planning and management in Accra. This approach is being initiated by the UNCHS in the fulfilment of the implementation of part of Agenda 21 of the Rio Earth Summit.

Environmental deterioration, an inevitable consequence of rapid urban growth, is caused in Accra, Ghana's capital, primarily by inadequate urban infrastructure and services, haphazard development and urban growth patterns that do not account for the opportunities and constraints of the natural environment. These short-comings have resulted, on the one hand, from lack of resources and insufficient investment in infrastructure and, on the other hand, from difficulties experienced by central and local government to effectively plan, co-ordinate and manage Accra's growth and development.

The environmental sanitation problem has constituted a major drawback to the economic growth and urban productivity of Accra. Liquid and solid waste have been poorly managed. The beaches of the city, recreational areas, markets and business areas are littered with refuse and garbage. The protracted problem of the Korle Lagoon and annual flooding in the city are all associated with the environmental degradation (The Consortium, 1994).

The UNCHS/UNDP supported the Town and Country Planning Department to prepare a comprehensive Strategic Development Plan for the Greater Accra Metropolitan Area in 1992. The primary thrust of the Strategic Plan is directed towards strengthening the economic base of the metropolitan area through strategies designed to increase industrial output,

171

improve land and housing delivery, strengthen commercial business and services and rehabilitate existing infrastructure and services. Emphasis is also given to improving the efficiency of public administration, planning and co-ordination, revenue collection and social well being (TCPD et al, 1992). The Accra Sustainable Programme (ASP) is intended to be used as a mechanism in the continuous execution of the strategies contained in the Strategic Plan. It will formulate action plans and projects on development issues that were identified in the Strategic Plan. These projects will then be implemented by the main line agencies and institutions.

The Working Group Concept

Actualisation of the concept and principles of the Sustainable Cities Programme is anchored in the Working Group, which is the basic building block of the Programme. Through working groups, partnerships are built between local governments, central government, councillors, parastatal organisations, the private sector, non-governmental organisations, community based organisation and individuals so that they may together formulate strategies and action plans to improve the long term environmental management in the city.

A working group is a mechanism for representing the key stakeholders in an operationally feasible manner. Through it, participation of the stakeholders is ensured in resolving the issues, and therefore a working group is expected to be empowered, motivated, transparent and action oriented (UNCHS, 1994b). The working group process does not supplant, but rather supplements existing institutions. Unlike traditional committees, working groups are believed to make connections of a cross-cutting nature. A working group is cross-sectoral in representation. It acts as a platform for negotiations, including negotiations on the political, managerial and implementation levels. It brings together institutions and forges partnerships. It can draw on the resources and expertise of the participating and sponsoring institutions. Consequently, it is able to transform its agreements into actions. At the same time, it is flexible enough to accommodate new stakeholders and to address new issues as they evolve. A working group is able to articulate visions and formulate strategies. It is able to focus on concerns of particular localities.

It must be noted here that almost all cities recognise the merits of participation and co-ordination. The notions of participation, governance and stakeholders have already become ubiquitous. The SCP therefore has not invented a new concept, but has provided a carefully designed framework and simple tools which allow it to consolidate and systematically apply participatory approaches to decision-making in urban and environmental management (Ibid.). In doing so the SCP introduces working groups as a cross-sectoral participatory mechanism with which to:

- prioritise environment and development issues;
- develop issue or sector specific strategies;
- identify, plan and implement demonstration projects in their communities and neighbourhoods;
- formulate sectoral action plans reflecting the issue specific strategies and for the replication of those demonstration projects and for the on-going maintenance and operation of those projects; and
- make inputs into the development of integrated multi-sectoral actions plans and an overall urban development strategy for the city.

Application of the Working Group Concept in Accra

The concept of a systematic environmental planning and management process formulated by the Sustainable Cities Programme provided the method being used in the Accra Sustainable Programme (ASP). This process aims to clarify environmental issues to be addressed, involve those whose co-operation is required, set priorities, negotiate issues specific strategies, co-ordinate overall environmental management strategies, agree on environmental action plans, initiate priority projects and programmes, and strengthen local planning and management capacity.

The working group concept was introduced in Accra through the ASP. The objective of the ASP is to support environmentally sustainable growth of Accra by strengthening the capacity of the Accra Metropolitan Assembly to plan, co-ordinate and manage environment and development interactions through using the working group concept. It is also aimed at creating an implementation capacity by acting upon a matter of high priority, in this case the complex of causes for the severe pollution of the Korle Lagoon. This involves issues of the management of both solid and liquid waste in the settlements along the streams feeding the lagoon, discharge of industrial waste into the streams and general environmental sanitation in the catchment of the lagoon.

The working groups are expected to produce clearer definitions of the problems to be addressed. This will provide more information with which to justify action and to obtain resources. Using this, they should then propose specific strategies for interventions that will reduce environmental problems associated with the degradation of the Lagoon. Following these strategies, they must produce detailed action plans to address specific aspects of each priority   environmental issue. In addition to spatial plans, these will include financial plans identifying capital investment projects, as well as operation and maintenance costs of these projects. There also must be institutional plans identifying which department, organisation, or agency is responsible for what activity in the implementation of the action plan and administrative and legal requirements.

It is intended that the working groups will consolidate their proposals into a single strategic development plan for the catchment area of the Korle Lagoon, embodying land, economic, and social development proposals, plus financial and institutional proposals. Part of this strategic development planning will provide an agreed review mechanism for evaluating the success of planning and managing growth and development in Accra. In the long term, The ASP hopes to use the lessons of its experience to implement the Strategic Plan for the Greater Accra Metropolitan Area, and then to replicate this approach in other urban centres of Ghana (UNCHS, 1994a).

In order to move through these stages, the city consultation first prioritised the most important environmental problems that should be addressed today to make the best use of the scarce resources available, since one cannot solve all the problems at once. This was done through a city-wide consultation, plus mini-consultations on subject-specific or geographic-specific issues.   Then, it built partnerships by bringing interested people and representatives from different agencies together into working groups so they might together prepare proposals to resolve the most pressing of those problems. The membership to these groups was drawn from the affected sectors and levels of government, geographic locations, the private sector, non-governmental organisations community groups and individuals. Then, it used the working group meetings to obtain information, ideas, skilled human and financial resources from the different agencies to implement the agreed plans. Such meetings also provide opportunities to resolve conflicting interests between institutions and groups. Furthermore the participation of individuals from the different institutions develops a sense of ownership of the ideas, and thus they have a committed interest in ensuring the sustained implementation of the plan proposals through their own existing public, private and popular sector institutions.

It is envisaged that the experience gained in the implementation of the working group process will be replicated in other parts of the city.

The Accra City Consultation on Environmental issues was held in May 1995. It was attended by Ministers of State, public servants, officials from Accra Metropolitan Authority led by the Chief Executive, traditional authorities, CBOs, NGOs, traders, private sector operators, research institutions, UN agencies and the general public. The environmental concerns for sustainable growth of Accra were associated during the consultation with the congestion in markets and lorry parks, informal manufacturing and small scale-industrial activities, obsolescence of service systems and congestion in the old parts of the city including the city centre, pollution and decay of the Korle Lagoon, gross inadequacies in health care and educational facilities, indiscriminate sand and stone winning for constructional works, and coastal erosion.

The consultation culminated in the Accra Declaration which called for follow up action to be taken regarding the environmental issues prioritised at the consultation. The consultation recommended that working groups be established to take up the following two overall priority matters:

- Sanitation, focusing on solid and liquid waste management, environmental health, and flooding and drainage.
- The Korle Lagoon, focusing on the pollution and degradation of the Korle Lagoon and the resource development of its catchment area.

It further recommended that additional working groups be established through mini-consultation processes to address adverse environmental conditions associated with informal sector manufacturing and services; congestion in markets and lorry parks; and sand winning, quarrying and coastal erosion.

To deal with the top priority issues plus the sub-issues in three geographic locations, six cross-cutting working groups have been formed:

- solid waste collection and transportation;

- solid waste disposal and recycling;
- liquid waste management and drainage;
- public education and awareness on environmental sanitation;
- pollution and degradation of the Korle Lagoon; and
- resource development of the Korle Lagoon.

Members of the working groups were drawn from those who are affected by the problem, those who can contribute expertise to the solution of the problem and those who possess implementation mechanisms to solve the problem. Some of these were identified during the city consultation as individuals prepared to serve on the working groups. These include people from central government departments and institutions, local government department and institutions, private sector, non-governmental organisations, community-based organisations, opinion leaders, traditional leaders and individuals. On the average a working group is composed of about ten people. The ASP carried out the selection of the working group members.

The three geographic locations were chosen to be pilot locations for the implementation of the Accra Sustainable Programme and all are in the catchment area of the Korle Lagoon. They are: a typical low income, high density residential settlement, called Ga Mashie; a large market for food stuffs and other manufactured products, called Agbogbloshi; and a small scale informal sector mechanics' area called Odawna. The total population of these areas is approximately 200,000. The types of solid waste generated by activities in the three locations are different. The first area generates domestic waste, the second generates market waste while the third generates waste of an industrial nature, such as used oil and scrap metal. The activities carried out in these areas impinge heavily on the Lagoon since they are all located in the vicinity of the lagoon and its tributaries. The activities thus contribute very much to the pollution and degradation of the lagoonal waters.

Results Achieved through the Working Groups

After two years, the ASP so far has prioritised environmental issues and established the six inter-agency and cross-cutting working groups to address them. This has achieved among many public agencies involved in the working groups an appreciation of the usefulness of engaging as many stakeholders as possible in finding solutions for urban environmental problems, creating through education in the mass media and involvement of the communities significant public awareness on environmental issues affecting the city residents.

The programme has conducted clean-up exercises with market groups which have led to a sustained clean environment in the market. Based upon a new awareness among market staff of the importance of market cleanliness, a partnership was built between the market groups and public sector agencies for the delivery of solid waste management services.

Analyses have been completed which determine what each stakeholder stands to lose with poor environmental conditions in the pilot areas. These use data collected on the pollution of the Korle Lagoon and the types of effluent discharged into the lagoon by the industries located along it. The result is intended to be used as a platform for negotiation contributions and commitment from the various stakeholders towards solving the problems identified.

The use of waste as a resource in composting, recycling, biogas production and fodder for goats and sheep has been studied. A major innovation in waste management has been implemented. A system of house to house collection of waste in the high density areas was set up. Following a strategy formulated by one of the working groups, young men and women are paid by households to collect domestic waste using donkey carts and then take it to a container site. This reduces unauthorised dumping on vacant plots and in open drains. And finally, a school sanitation education project has been proposed, to be financed by UNICEF.

Lessons Learned

The Accra Sustainable Programme is gradually becoming a focal point for interested individuals and institutions to identify with and become involved in managing the city's environmental resources. UN agencies and the donor community find it easier to channel funds into the projects which are well prepared and which involve as many stakeholders as possible.

In spite of recognition of the relevance of the working groups, the ASP has faced a number of problems. The major ones begin with the fact that many administrators and politicians consider the ASP to be one of those donor projects which bring in money for construction of projects rather than a process of doing the same things differently. Then, because professionals tend to be trained conventionally, substantial time is required to sensitise and indeed "de-school" various actors who participate in the working groups to respect and adopt to other professional views and individual opinions. Time is also needed to introduce professionals, technocrats, decision makers, members of the public, individuals, etc. to the project principles, process and activities of the Sustainable Cities Programme approach.

Because the ASP focuses on environmental issues specifically to cut across institutional divides, so certain institutions, departments, professions and indeed individuals feel threatened. Fears have been expressed of a take-over of other City departmental functions by the City Planning Department, where the Technical Support Unit of the ASP is located. This has required more sensitisation of these other departments to the Programme principles and process, emphasising that the Programme is there to support all institutions involved in better management of city resources and is not itself a new department.

Perhaps of greatest concern is that there is constant pressure to demonstrate progress to keep actors, especially decision makers, motivated and involved. Yet the ASP is essentially a long term process which delivers visible results gradually and whose most important achievements are institutional changes and attitudes. It takes longer to institutionalise a process than to set up a blueprint, yet stakeholders expect so much from the ASP within a short time.

Conclusions

The systematic environmental planning and management process initiated in the ASP seeks to build partnerships with all stakeholders through working groups process in order to resolve urban environmental problems. The process builds up gradually and takes time to deliver visible physical results. It creates conditions which may more easily attract donor assistance for bankable investment packages, so many administrators expect the ASP to bring money to implement projects as other donor-driven programmes. However, the ASP is a process for the preparation of investment projects through participatory mechanism which brings together all those who are connected with the problem. The ASP then takes up the project proposal to solicit funding both internally and externally to implement it. Like all others in the Sustainable Cities Programme, the ASP does not have implementation money of its own.

The process advocated by the Sustainable Cities Programme is ultimately not dependent upon international intervention. It brings together government departments, institutions, groups and individuals to plan and implement projects. It builds capacity for local governments to plan. Viewed in this light, the process is sustainable if properly nurtured. The Accra Sustainable Programme is still in the donor-assisted stage, but the evidence to date is that, after the assistance is withdrawn, the process will be part of the normal working practice of the Accra Metropolitan Assembly.

References

The Consortium (1994), 'An Environmental Profile of Accra Metropolitan Area', Accra (July).

TCPD (Town and Country Planning Department), UNCHS, and UNDP (1992), *The Strategic Plan for the Greater Accra Metropolitan Area*, Volumes 1 and 2, Accra.

UNCHS (1994a), 'Managing Sustainable Growth and Development in Accra', Project Document: GHA/DN/GLO/2/D.12, Accra.

UNCHS (1994b), *Sustainable Cities: Concepts and Applications of a United Nations Programme*, UNCHS, Nairobi.

16 The Development of a Systematic Approach to Urban Environmental Planning and Management in Thailand

ADRIAN ATKINSON

The Urban Environmental Management Project (UEMP) originated out of the problems that had arisen within a number of GTZ-funded urban projects around the world. On the one hand, solutions to urban problems were being addressed in a fragmentary way, both because of the uneven availability of information and of the sectoral and technocratic approach to management. Not only residents of poor communities, but also local politicians, generally possess only a very partial knowledge of the structure of environmental problems or the range of options available for solving these.

On the other hand, the decision-making process regarding urban management is also fragmented as between various departments, levels of government, 'inside' interests such as business organisations, and other interests, including the organisations of poor communities. However well a poor community gets its act together to manage local environmental problems, without the active cooperation of other urban groups and, in particular, the local authority, net benefits may be questionable and disillusionment is likely to follow. The famous Orangi project in Karachi is an excellent case in point, where the hostility and hence the lack of cooperation of the local authority meant that the self-built sewer system actually exacerbated water pollution in other parts of the city.

The UEMP was created to pioneer new methods. It was clear that it would be necessary to work simultaneously at the national, city and community levels, which would mean running a series of articulated sub-projects, some of which would be structural whilst others would demonstrate in detail how the system would work, particularly at the all-important community level.

The UEMP in Thailand

The project in Thailand was initiated by a three-day workshop involving representatives of key national agencies, a selection of local authority representatives, NGOs working in the urban field and academics working on urban environmental management issues. The workshop identified the key environmental problems in Thai cities, laid the foundation for a support network, and established an institutional base in the Office of Urban Development (OUD) of the Department of Local Administration, of the Ministry of the Interior.

For the first few months, the project produced a series of pamphlets on the environmental issues identified in the initial workshop. These included "brown agenda" issues of water pollution, solid and hazardous waste, air pollution and environmental health, as well as more complex issues of transportation, the built environment, land management, organisation amongst poor communities and even regional resources management.

The pamphlets were not aimed at technicians, but at providing technical information to key local decision-makers including mayors and councillors, City Clerks and key non-municipal actors interested in becoming involved in local development and management decisions. The approach - the style - of the pamphlets was therefore to ask simple questions and to provide easily-understood, and liberally illustrated, answers and options. Case examples were provided to illustrate what had already been done, and each pamphlet contained a directory of sources of assistance. The emphasis in this directory was on networking between communities and local authorities, rather than calling upon a higher authority.

Even among the consultants hired to produce the pamphlets, there were problems with thinking in terms of lay understanding and simple practical solutions: professionals are used to reinforcing the elite decision-making process through esoteric approaches to urban management and infrastructure provision. It took considerable work to convert what the professionals provided in terms of materials into guidelines genuinely accessible to non-professionals. Moreover, it became clear in the course of the project that emphasis was needed upon the provision of specific technical and organisational information to non-government and community groups and a second set of guidelines were produced to satisfy this need.

Meanwhile, seven municipalities, ranging from Chiang Mai with a population of over 200,000 people to Paak Phanang with less than 8,000, were chosen in which to operate the project, not only because of the variety of their key problems (coastal, inland, tourist, industrial, etc.) and regional scatter, but primarily because of the probability of their making good progress, as indicated by strong commitments from the politicians and/or key officers, and the existence of some NGO and/or CBO activity.

A set of procedures for the institution of local environmental action planning were drafted and, together with the Guidelines, were critiqued by all the participants. Then over a six month period in late 1992 and early 1993, the seven municipalities underwent a pilot (in Thai:, *Nam Rong*) process of generating environmental action plans. This involved the following:

- Each municipality organised an environment committee comprising municipal politicians and officers, representatives of other local interests and representatives of the provinces (which, in Thailand, possess powers and budgets to carry out urban works); each municipality also appointed a liaison officer who was responsible for monitoring progress for the project.

- Following the prioritisation of issues, working committees (including non-government participants) used the Guidelines to consider what approach should be taken to solving the prioritised problems.

- Meanwhile, certain generic issues were identified, such as solid waste landfill site construction and management, and methods of municipal-CBO cooperation, and workshops were held on them. These workshops involved participants from all the Nam Rong municipalities, in order to initiate demonstration projects that would produce solutions that might be emulated elsewhere.

- A general process of awareness-raising was encouraged, both to instil a greater concern for environmental problems and to make the connection in the minds of the public between these problems and the concurrent action planning process. Events ranged from local conferences to a whole week of events: for example, a well-known Buddhist monk delivered a sermon to several thousand and a shadow puppet play was written and staged.

The procedures made use of existing skills in the local authority in assembling municipal budgets, and could even be interpreted as a 'popularisation' of the conventional budgetary process. The formal plans also called for non-government actors and resources to participate in developing and implementing detailed solutions. Whilst this fell far short of a fully rounded environmental planning exercise, it did provide a foundation upon which further rounds of training and planning could be built, and it had the advantage of starting from familiar ground.

The level of activity was very different from one municipality to another, with some showing little by way of results but, on the whole, with a wealth of useful experience being accumulated. After the Nam Rong exercise was formally concluded in April 1993, further work was done on refining the Guidelines, which were eventually published and distributed. Although the UEMP did not have the resources to follow up on all the plans produced by the municipalities, some demonstration projects were undertaken.

Some Background Considerations

Although local government is weak in Thailand, the central government, in the form of the Department of Public Works (DPW), does, in fact, spend considerable resources in providing infrastructure directly - urban roads, river protection, water treatment facilities, etc. The snag is that this does not necessarily represent what the local authority, and in particularly local communities, consider to be priorities or appropriate uses of resources. In the end, there is no feeling of 'local ownership' of the end product, which is consequently badly managed and perfunctorily maintained.

In this context, the Nam Rong process was designed to initiate a new kind of local politics of urban management that would have local government and communities making common cause concerning what they see as priorities, and estimating the resources (including personnel capabilities) necessary to carry out effective local environmental management. But, given the inadequacy of local finances to cover larger cost items, how could demands for the necessary resources be articulated effectively at the national level?

The first hurdle was to greatly improve the inclination and capacity of local authorities and communities to work together to create a 'united front'. Legally and according to tradition, local government in Thailand is, in fact, little more than an arm of central government and therefore basically uninterested in gaining greater local autonomy. Local NGOs and CBOs know this and this creates a hostility which both confirms local government in its detachment from local commitment and reduces the possibilities for real community benefits arising from local government programmes. The UEMP thus worked very hard to forge a working relationship between community organisations and local government in order to capture local government as an ally in the struggle to gain more powers and resources from the centre.

Two approaches presented themselves. The first recognised that, because national government bureaucracy is always the scene of struggles for control over resources (Riggs, 1966), there are usually opportunities for new actors to appeal to the interests of particular agencies in the thick of the competition. De facto decentralisation through the partial measures may be achieved this way, aided by the fact that decentralisation in Thailand is currently seen in general political terms as 'a good thing', particularly within the ideological context of environmental management. For example, the passing of the Environmental Quality Promotion Act in 1992 established a substantial Environment Fund, controlled by central government. Hitherto, the DPW had been responsible for delivering environmental infrastructure as finished objects: urban roads, sewage treatment plants, solid waste landfill sites, and so on. Under the new Act, local authorities may now formulate their own solutions to water pollution and solid waste management problems and bid for money from the Environment Fund to implement their own solutions, including not only hardware but also institution-building and public involvement.

There had been serious problems in implementing this mechanism - not least of all, the lack of local capacity to formulate proposals and generally plan and manage the process. Consequently, the UEMP intervened both at the local level, by way of demonstration projects, and at the national level, with a view to tying the procedures for application for money from the Environment Fund into the urban environmental action planning procedures which it was developing.

The second approach was made via the Municipal League, the association through which municipalities look after their collective interests in the political arena, which until recently had been little more than an arm of the Ministry of Interior. From the start of the project, the UEMP targeted the annual meeting of the League as a venue to disseminate both the environmental management and the decentralist agendas.

This paid very big dividends. From the outset, the Mayor of Phuket, one of the Nam Rong municipalities, embraced the project wholeheartedly. The municipality thus became a model for the action planning process, demonstrating cooperation between the municipality and NGOs in place of what had been a deep-seated suspicion, if not outright hostility. When the Mayor became the Chair of the Municipal League for the year 1994-95, he created the elements of a national 'Local Agenda 21' process on advice from the UEMP. By networking with other Mayors attracted by this approach, he established a Health and Environment Committee to develop the League's Local Agenda 21 programme. The League spent its own funds to republish the UEMP Guidelines and to develop and circulate further materials. A series of regional workshops are being organised by the committee to raise awareness and disseminate information and methods for urban environmental action planning.

Shortcomings of the Project

Of course an experiment such as this must be expected to have shortcomings - indeed some of the most constructive lessons grow out of failure. It is therefore useful to look at some of the shortcomings of the UEMP.

Chiang Mai

Not all the Nam Rong municipalities made the running. The failure of Chiang Mai, Thailand's second city, was dramatic. Chosen because the City Clerk was both respected and keenly interested in participating (he was involved in the initial Project workshop), and because of the presence in the city of very strong local NGOs (Rüland and Ladavalya, 1993), the attempts to operationalise the environmental action planning process were nevertheless effectively blocked by the Mayor.

What was particular disheartening was the way in which the Mayor and his associates had created a machine politics amongst the poorer communities designed to assure his continued re-election, whilst running an openly corrupt system of urban management. Rampant construction of (largely vacant) high-rise buildings has been encouraged within a completely inadequate planning framework. Meanwhile, in spite of attempts of the UEMP (and various other donor agency projects) to assist in the introduction of an effective waste disposal system, solid waste has been dumped across the surrounding landscape in a completely unregulated manner, arousing local demonstrations and emerging as a national scandal, but continuing to go unsolved. In the autumn of 1995, however, the Mayor was voted out and it seemed that there might now be possibilities for progress to be made by local NGOs who have been trained in the principles and the procedures developed by the UEMP. Unfortunately these hopes for change were soon dashed.

Paak Phanang

The small municipality of Paak Phanang undertook to implement a demonstration project designed to create an appropriate and effective waste disposal system, including recycling and a properly constructed and managed sanitary landfill site. The UEMP supplied the municipality with expert advice on how to put together an appropriate project and also assisted in the writing of a successful bid for funding from the Environment Fund to pay for engineering and other assistance to develop the project.

The UEMP was unable to assist at each stage of the process which ensued, with the result that the municipality obtained the same kind of disempowering and indigestible expert report as characterised 'technical assistance' before the advent of the UEMP. The awareness-raising and community participation that were supposed to characterise the exercise failed to materialise.

Nong Khai

One of the outputs of the Nam Rong environmental action planning exercise in Nong Khai was a recognition of the need for effective land use planning. Around this the UEMP organised a demonstration project which involved participation by the municipal staff and community representatives. With advisors, they devised procedures for bottom-up land use planning and applied them, going against the normal procedure wherein the central government Department of Town and Country Planning performs land use planning. This exercise in Nong Khai was a complete success in that the communities and municipality are now capable of applying basic land use planning techniques. However, drawing up a plan is the least of problems. Afterwards, both sides hesitated to act: the local community, including the land owners, were loath to honour the plan - which would effectively require giving up land for ample infrastructure rather than selling off as much as possible to house builders - unless the local authority demonstrated its willingness to pay for the infrastructure. On their part, the municipality asserted that they had not been able to convince the DPW to put this infrastructure into their investment plans.

Apparently the DPW were very open to suggestions from the municipality concerning where to build infrastructure; furthermore, special funds have been allocated to border towns, including Nong Khai, specifically to respond to increasing development pressures. At the local level, however, the perspective was somewhat different. Because supplicants to the DPW from local authorities do not get the attention accorded to the Bangkok elite, and they are more timid in making their applications, they may decide simply to avoid making difficult requests. Furthermore, local elected officials working with small budgets have to make very fine balances by way of 'paying off' each community in order to assure their political survival and so are not necessarily happy suddenly to favour one community just because it has a plan to demonstrate its need for additional infrastructure funding.

Although all these shortcomings do not denote failure of the demonstration projects, they certainly indicate that such projects require longer-term nurturing and intervention at many levels to ensure that they progress to a successful conclusion. In essence, the problem involves one of developing trust between local authority and community, that has to prove justified in practice, and overcoming the "culture of subservience" that pervades local-central government relations.

The State of Play

Some municipalities, particularly those with Mayors who have a keen interest in the project, are taking up environmental management in earnest and making use of the training materials produced by the UEMP. Prepared by the project, a group of trainers, including OUD staff, senior municipal staff, academics and NGO personnel, are available through the Municipal

League to carry out local training activities where there is a demand.

The OUD staff have become genuinely committed to the UEMP. In the early stages, changes in staff and the unfamiliarity of the objectives of the project meant that collaboration was somewhat perfunctory. Then OUD staff stabilised, and they have been in position long enough to internalise the project aims and discover the advantages to the OUD of providing positive support. After the OUD proposed that the environmental planning and management methods of the Nam Rong process be disseminated throughout the municipalities, Government in 1995 required that these procedures be implemented through a Ministry of the Interior Regulation as a standard process for determining the content of the municipal budget.

GTZ was invited to help build the experience, and hence the procedures, into the 8th National Social and Economic Development Plan. This National Plan is intended to be a major departure from earlier Plans, which were essentially concerned - and with notable success - with achieving economic growth at whatever cost. Growing criticism of the inadequate concern for the social and environmental costs of this strategy have precipitated a proposed major change in direction for the new plan to pursue a path of human sustainable development that puts social and environmental goals to the fore. Unlike previous plans, production of this plan has also involved an extensive consultative exercise, with widespread debate especially concerning the role of non-government and community-based organisations in the urban decision-making process.

Conclusion

A project in Thailand, supported by the German Agency for Technical Cooperation (GTZ), which ran from early 1991 to the end of 1994, attempted to address the full spectrum of issues regarding the effectiveness of a participatory approach to urban environmental management and planning. Focusing on the relationship between local community problems, aspirations and capacities, and the roles, functions and capacities of local government, the project developed a set of general urban environmental planning and management procedures that both enable local participation and self-activity and provide a basis for government at various levels to respond in enabling ways. This was subsequently enacted as the standard method for municipal planning - albeit there remains much work to be done before it is properly understood and practised throughout urban Thailand.

Note

This paper is an updated and much edited version of Atkinson, A. and Vorvatuchaiphan, C. P, 'A Systematic Approach to Urban Environmental Planning and Management: Project Report from Thailand', *Environment and Urbanization*, vol. 8, no. 1, 1996 (April).

References

Riggs, F. W. (1966), *Thailand: the Modernisation of a Bureaucratic Polity,* East-West Centre Press, Honolulu, USA.
Rüland, J. and Ladavalya, M. L. B. (1993), *Local Associations and Municipal Government in Thailand*, Arnold Bergstreasser Institut, Freiburg i.Br, Germany.

PART III
ORGANISATION
AND POLITICS

17 Organisation and Politics in Urban Environmental Management

JULIO D. DÁVILA AND ADRIAN ATKINSON

This section groups together eight papers by academics and practitioners writing about a number of countries in five continents. Despite the geographical and disciplinary diversity of their authors, these essays have in common a concern with the political and organisational context in which urban environmental management takes place. The aim of this introductory chapter is to examine a range of political and organisational factors that help shape the practice of environmental management in cities.

A starting point of the chapter is the premise that, insofar as political and institutional (or organisational) realities vary from one geographical and historical context to another, an adequate understanding of the national and international context in which urban environmental management takes place is important. Three global processes running in parallel are mentioned which help define the framework within which national and local politics and institutions function. The chapter then identifies a series of factors operating nationally or locally which also help influence the practice of urban environmental management. These include the issues of institutional responsibilities and political leadership, the scale at which action takes (or ought to take) place, the issue of changing attitudes, the new sets of organisational skills needed to respond to changing political realities, and the role of external support agencies.

The Global Context of Urban Environmental Management

While the main concern of this book is with the management of the urban environment, one must recognise that all forms of management occur within a specific economic, cultural and political context. Management offers practitioners a set of tools and a range of resources to achieve their objectives. Such tools and resources vary from one economic, cultural and geographical setting to another and actions to manage the environment arise from the specific circumstances at hand. While there are principles of good practice in environmental management, the very fact that different societies (and groups within each society) perceive their own environment differently suggests that there are no universal recipes for tackling urban environmental problems world-wide. The political will to carry out a set of actions and the institutions that might undertake such actions are unique to that historical and geographical context (Younge, 1997; Burton, 1997).

At the end of the twentieth century the global context is marked by three major features. The first one refers to major global environmental change largely resulting from human actions. The consequences of these actions are relatively well documented and range from the regular human use of more than half of all accessible surface fresh water, to increases of nearly a third of carbon dioxide in the atmosphere since around 1800, to the elimination of about one-quarter of all bird species on Earth (Lubchenco, 1998). This has significant implications for policies towards the processes of territorial expansion of metropolitan areas now so widespread around the world (Monclús, 1998; UNCHS, 1996).

The second is an increased integration of the world economy and society to a scale and at a pace never before seen in history. This integration (known by the often poorly-defined term of globalisation) takes shape in a whole range of dimensions of human life: cultural, economic, in the way work is organised, in communication techniques and so on (Beck, 1998). The world economy is increasingly supported on a reduced number of world cities where the majority of financial transactions take place every day aided by a network of secondary cities (where much smaller volumes of global business are conducted but which serve an important national or regional purpose), while other cities and their inhabitants are largely by-passed by many of these processes of global integration and their accompanying consumption patterns.

And the third is a substantial redefinition of the role of the state and a greater and changed role for other societal actors. In many nations, the urgency of political realities has impelled the state to become more sensitive and responsive to the demands placed by its citizenry particularly regarding their local environment. Reform has often involved a dual process of decentralisation and democratisation and in consequence an increased importance of local and regional governments in the political conduct of national affairs.

The Relevance of Global Trends to Urban Environmental Management

The fact that many environmental problems (such as oil spills, acid rain, global warming or nuclear accidents) are no longer restricted to circumscribed localities but have regional or even global consequences, means that localities cannot afford to isolate themselves and pretend to ignore the effect distant incidents might have on them. Local authorities and citizens in what might appear to be isolated localities need to be aware of global environmental trends and need to seek to influence processes that might affect them or to control locally-originated phenomena that might affect others, however distant they might appear to be. And yet, the notion of ecological globalisation also serves to highlight the growing impotence of the national state vis-à-vis global integration. As Beck (1998) has remarked, 'globalisation means... absence of a global state... a world society without a world government' (p. 32). Thus, despite the apparent relevance given in the international media to treaties and agreements such as those discussed in the 1992 UNCED conference in Rio de Janeiro, it is disheartening to see the poor enforcement of such treaties and agreements and the lukewarm response of ensuing international meetings (such as that in New York in 1997 or Kyoto's global warming meeting in December of the same year).

Although, as the paper by Marvin and Guy included in this section argues, the local level is not necessarily the best level for environmental action, it is perhaps only by generating a critical mass of local actions that address global environmental problems that the national state or supranational associations such as the European Union might hope to influence global environmental politics. Insofar as the state of the environment affects everyone's present and future it cannot be left to market forces. It needs concerted actions by a group of stakeholders and other actors (i.e. those who might not be affected directly, at least in the short run) who might seek to exert some influence in the limited space of their locality but seeking to project their actions at the global level (a case of 'act locally, act globally').

The reformed state described above might not appear to be the ideal vehicle to channel such kinds of environmental actions. It might seem too weak vis-à-vis transnational capital, or too over-burdened with the fiscal and political weight of pensions, the army and a growing number of poor people and social inequalities. And yet, the state is still in a privileged (and irreplaceable) position to co-ordinate the environmental action of citizens, to penalise and control large polluters, and to reform its own institutions so they become more responsive to environmental demands. After all, the state still has a legitimate (though not undisputed) monopoly over weapons of war and repression, and over legal instruments affecting the lives of all citizens. Although not perfect by any means, it is still the best mechanism we know to help alleviate poverty at a significant scale, re-distribute income and promote greater social equity within national boundaries.

State reforms in the direction of greater democratisation and decentralisation represent an opportunity for dealing more successfully with urban environmental management. The state cannot be the only body responsible for looking after the environment. As many have emphasised in successive international conferences and countless documents, the environment is a shared responsibility, too important for the well-being of present and future generations to be left to any one social actor. As Souza remarks in her paper, the dilemma for politicians who are responsible for running the state, including the local state, is that the environment rarely generates votes, especially in the poorer countries of the South, as it tends to involve long-term and not immediately visible efforts.

As several papers in this section argue, the challenge lies in involving other actors in the process of looking after the environment or formulating actions and strategies for doing so in the future. In some cases, as Church shows for the United Kingdom, local authorities play a crucial role in initiating environmental actions. And sometimes, as Yonder demonstrates for Istanbul, the influence of actors such as a professional association becomes decisive in counter-acting flawed and corrupt decisions taken by national, or indeed local, politicians. In the case described by Velásquez and Pacheco, the role of actors different from the local authority was very important to help diffuse a situation of environmental conflict.

Decentralisation (understood as transfer of significant powers and resources from central to regional and local authorities) opens the door to more flexible and effective action to deal with the environment at the local and regional level. As Allen shows in her paper, the powers and autonomy given to local authorities to create networks and join institutional forces has had beneficial results for the environment in Latin America. It has not, however, been so beneficial in the case of neighbouring local authorities where one is perceived as exerting its power at the expense of others, as Velásquez and Pacheco show for the Manizales-Villamaría conurbation.

But decentralisation reforms have not as yet helped to solve institutional impasses such as those resulting from institutional overlaps or lack of clarity in terms of the duties and responsibilities of different agencies. A frequent case is found among watershed management agencies which often enter into conflict with municipal or regional governments over the generation or enforcement of regulations within their territory. The conflict appears to be sharper in those cases when watershed managers are appointed by a higher authority while local or regional heads of government are elected by popular vote.

In those national instances where central bodies still retain prerogatives over the generation and enforcement of common standards and procedures, such as the case of flood management described for New South Wales (Australia) by Handmer, persuasion and dialogue appear to be more effective than coercive measures. It has also been remarked that environmental legislation imposed from above cannot succeed unless it is matched by grassroots development of civic consciousness which relate environmental issues to people's own concerns (Burton, 1997).

As Mattingly suggests elsewhere in this book, there are few experiences around the world where the urban environment is managed as an integrated whole rather than as an add-on to existing local government functions and institutions. This raises a number of issues regarding the organisation and the politics involved in urban environmental management.

Institutional Responsibilities and Political Leadership

The first of these issues refers to the institutions responsible for managing the environment and the actors involved. At the level of local or regional governments, responsibilities are rarely defined along lines other than sectoral ones. In those contexts where an institution has actually been given the responsibility for managing the urban environment, it has tended to become yet another sector so that important links across sectors are rarely made. The danger is that municipal line secretariats can often afford not to incorporate environmental (much less sustainability) concerns within their actions with the excuse that these are the business of environmental agencies. What is needed at the local level is a strong institution that can take the leadership of environmental issues so sustainability alliances can be built across sectoral divides.

Leadership is also essential in the frequent case when environmental conflicts arise (Younge, 1997). This is especially true in those cases where there is a mismatch between the territories of the environmental authority and the water catchment area, as is the case of Istanbul examined in Yonder's paper. Leadership may also ensure that the voice of communities claiming some form of ownership of the commons (e.g. water resources or land) is taken into account in the decision-making process (Burton, 1997).

Supra-municipal leadership may also help reduce regional or metropolitan conflict resulting from 'environmental racism'. Conflict might not arise in cases when environmental problems (such as solid waste disposal) are shifted to localities where costs, local resistance and political and economic power are lowest but in those instances when less powerful groups organise and resist the displacement of environmental impacts to them. Even if initiatives and leadership do not come from municipal institutions, these must recognise the importance of perceiving the environment as belonging equally to everyone not merely within the limited jurisdiction of the locality but also within a wider area which may involve other townships downstream or distant regions which provide the raw materials and even the finished products consumed in the locality.

Political leadership and vision become essential in incorporating actors who are not directly affected by environmental change into the dynamics of urban environmental management. But above all, leadership must in the medium-term seek new frameworks for greater participation both of stakeholders and non-stakeholders, thus planting the seeds for more effective future action.

Scale of Action

This leads us to the issue of the scale at which actions for sustainability ought to take place. The 'new localism' which Marvin and Guy argue permeates much current thinking on the environment, uncritically perceives the local scale as the ideal level of intervention and action. This perception has a number of shortcomings, not least the fact that many actions (such as regulation leading to new forms of production and consumption) are more effective if undertaken at the national level. As several of the papers in this book appear to suggest, within a given a set of limitations in terms of territorial jurisdiction and given limited fiscal and human resources, urban environmental management would appear to be best tackled at the largest possible scale (here the principle of subsidiarity would also apply, with the appropriate provisos regarding environmental priorities).

For example, as Allen shows for the associations of municipalities in Latin America, when the scale of action has increased, effectiveness has tended to increase as well. However, in other contexts, such as the UK, cuts in local government finances and increased fragmentation of local powers have militated against such a comprehensive approach.

In the case of metropolitan areas, there is evidence to suggest that there is no logical necessity for a special metropolitan level of government to be created to address environmental management at that scale. Action at the metropolitan level can be undertaken by any level of government, be it provincial or national, or even an association of local councils (Younge, 1997).

Attitudes

It would appear that there is widespread agreement at least among environmental specialists and activists that a change of attitude towards the proposals of urban environment management is needed. This is particularly true of top-down approaches to urban management and planning, which militate against much of the participation and empowerment discourse advocated even by multi-lateral development agencies like the World Bank. In fact, in many metropolitan contexts, although a preoccupation for the environment ranks highly among citizens' concerns, their views are rarely consulted.

However, when it comes to seeking a shift in attitudes, Marvin and Guy are critical of the homogenising ethic of the 'new localism'. They advocate targeting professional associations rather than all individuals in society for, they argue convincingly, individual perceptions are not moulded only by geography but by a multiplicity of factors. However, it is worth noting that although such as an approach might be effective in the richer societies of the North where virtually every member of society belongs to some sort of association by virtue of their occupation (even the unemployed are grouped together and looked after by the state), significant number of workers in poorer societies lack any kind of professional affiliation and cannot therefore be reached through such associations.

New Organisational Skills

Urban environmental management requires a new set of skills not merely on the part of government officials but, equally importantly, on the part of private businesses, professional associations, communities, NGOs and the staff of external support agencies. There is an urgent need for strengthening and building the capacity to act in new and more effective ways with the aims of environmental and social sustainability in mind. Thus, for example, the poor staffing and training levels of environmental monitoring agencies are part of the problem described by Yonder for Istanbul's water reservoirs. Similarly, the role that local universities and professionals can play in guiding urban environmental management is highlighted in the papers by Yonder, Velásquez and Pacheco, and Kyessi.

External Support Agencies

Finally, one must not overlook the potentially beneficial role that external support agencies (such as multi-lateral and bi-lateral aid institutions) can play in urban environmental management. As the paper by Kyessi shows for Dar es Salaam, external aid can sometimes play a crucial role in ameliorating poor environmental conditions while providing the necessary support to communities to help them take greater and more effective control of their lives. External agencies are in a privileged position not merely to provide financial support, but perhaps more importantly, to learn and adapt lessons from other contexts which might help in generating a more effective and sustainable approach to urban environmental management.

Conclusions

As the world becomes increasingly integrated economically, culturally and environmentally, there is a risk that elected politicians and other decision-makers may forgo some of their responsibilities with the (not entirely implausible) argument that there is little they can do to counteract the effect of global forces. Capital and with it jobs will inexorably move out of countries where it cannot get an adequate return. Pollution will be exported to those localities which exert the least resistance. A homogenising ethic to deal with environmental problems will override traditional ways of working in community.

This might well be the context in which urban environmental management increasingly takes place in the twenty-first century. And as the world population continues to urbanise, cities around the world will take on new and increased relevance not only as the foci of environmental conflicts, but also as the loci of individual and community action to face global environmental deterioration. Many of the papers discussed here describe positive experiences in dealing with environmental conflict and deterioration. Much of the hope that emanates from them stems precisely from the fact that the global trends described above have been used to the advantage of local populations with sustainability principles in mind.

In the context of a reformed state, a form of political leadership which can tap onto the energy of communities and the goodwill of the private sector is a decisive element in the process. It would also seem that, providing that there is a modicum of consensus about the need to overcome environmental problems, a sense of urgency in achieving greater social justice, and adequate mechanisms for the active participation of stakeholders and other actors, there is no need to wait for new institutions to be created to adequately confront environmental problems at a metropolitan or regional scale. Imagination and a common sense of purpose can play a decisive role in helping to shape local attitudes, in strengthening institutional capabilities and individual skills, and in tapping the resources of external support agencies for the benefit of poor communities and the global environment.

References

Beck, U. (1998), *¿Qué es la Globalización? Falacias del Globalismo, Respuestas a la Globalización*, Paidós, Barcelona.

Burton, J. (1997), 'Water management', workshop rapporteur's summary, Conference on *The Challenge of Environmental Management in Metropolitan Areas*, jointly promoted by Development Planning Unit and Institute of Commonwealth Studies, University of London, 19-20 June (mimeo).

Lubchenco, J. (1998), 'Entering the century of the environment: A new social contract for science', *Science*, vol. 279 (23 January), pp. 491-497.

Monclús, F. J. (ed), (1998), *La Ciudad Dispersa*, Centre de Cultura Contemporània, Barcelona.

UNCHS (Habitat) (1996), *An Urbanizing World. Global Report on Human Settlements 1996*, Oxford University Press, New York.

Younge, A. (1997), 'Institutional development and conflict resolution', workshop rapporteur's summary, Conference on *The Challenge of Environmental Management in Metropolitan Areas*, jointly promoted by Development Planning Unit and Institute of Commonwealth Studies, University of London, 19-20 June (mimeo).

18 Beyond the Myths of the New Environmental Localism

SIMON MARVIN AND SIMON GUY

Introduction

The environment of contemporary cities is currently taking centre stage in both advanced capitalist and developing countries (Haughton and Hunter, 1994; Pugh, 1996). Newspapers carry dire warnings about the 'impact' of urbanisation almost daily and we have become accustomed to gloomy predictions about the ecological future of cities. But for much of the 1990s academics and policy-makers in local government in the United Kingdom have been telling a much more comforting story. A distinctive new intellectual and programmatic discourse has come to dominate the urban environmental debate. We will call this the 'new localism'. The story goes that environmental policy initiatives at the local level will effectively deal with the ecological chaos of today by creating a more rational future with local government leading the development of more sustainable communities, life and work styles. In this way cities will regain political leverage by reconstructing a new form of transformative local governance around the environmental agenda.

The localist discourse is emerging as the main conceptual frame through which local environmental policy is viewed. We think this is worrying because the localist story rests on an incomplete analysis of what shapes cities and provides only a partial guide to urban environmental policy. In order to highlight these problems the paper critically examines six core myths that underpin the new localist orthodoxy. Some of the myths are more pronounced than others and some more critical to new localist thinking.

Of course not all the sources quoted in this paper are guilty of all the claims made by the new localism. In order to sustain our argument we have had to do some violence to the differences within the literature. Critically, our use of the word myth does not suggest that the discourse is a form of fiction or untruth. Instead the concept of a myth is used to represent narratives that concretely frame the way in which environmental problems are conventionally viewed. So we will begin by outlining what we feel are the main 'myths' underpinning new localist discourses before identifying what we miss by clinging to these myths.

Mythologies of the New Localism

Myth 1: The treatment of environmental problems needs to be tackled at a local level

The importance of the locality is often asserted as being pivotal and self evident in the shift to a more environmentally sustainable future. The central tenet is that the locale has unique features that justify the development of research programmes and policy initiatives at this level. It is assumed that individuals feel powerless in relation to spatially and socially distant environmental problems and that they relate to them, if at all, at only an abstract or ethical level with few direct consequences for action. In this context the key to changing individuals' everyday behaviour and consumption patterns is action through 'local initiatives and intercommunal processes' (Ward, 1996, p. 137). Conceptually the locale is seen as a socio-spatial container into which the sum of institutional, social and physical relations necessary to achieve a more sustainable future can be found. The local becomes a 'black box' disconnected from the global, international and national contexts within which localities are framed. Not surprisingly the local has become 'enshrined' in much of contemporary policy development (Agyeman and Evans, 1994, p. 11-12).

Myth 2: Local government is best suited to tackle these problems through the practice of the new local environmentalism

The new localism has provided a new logic around which local government is reinvigorating and transforming itself as a 'key player in the transition to sustainability' (Selman, 1996, p. 4). Under the stimulus of international and European environmental agreements the environment has developed into 'one of the most important policy areas' for local authorities who are now the 'leading actors and agencies' in environmental policy and practice in Britain (Agyeman and Evans, 1994, pp. 1 and 13). Environmentalism is entering almost every element of local government practice creating new discourses around green cities, conservation and ecology, the modelling and measuring of the environment, new institutional processes, the role of the environment in economic development policy, the local authority itself and new landuse strategies (see Hams, 1994; Webber, 1994). The uncertainty about what an environmental strategy actually constitutes is 'less important than the mushrooming growth' in local actions (Webber, 1994, p. 47). Local government is widely seen as having the key co-ordinating role in these multiple motifs that make up the new localism in environmental policy. Because environmental issues cut across traditional service boundaries the local authority becomes a 'catalyst' of community action building on its strong links with the public and its role as a 'local advocate and shaper of opinions' (Webber, 1994, p. 57). This agenda is 'transforming' local government as planners adopt new structures and new policy initiatives, think in more creative and original ways and encourage the involvement of community and business sector. These shifts have created 'a new vocabulary, a fresh set of principles, attitudes, theories and values' (Agyeman and Evans, 1994, pp. 7-8).

Myth 3: Citizens have to change their behaviour by becoming active participants in the implementation of the new localism

Here the assumption is that local citizens need to overcome barriers to changed forms of behaviour in order to develop more sustainable lifestyles. This is linked to the presumption that these shifts are best facilitated through local action which can highlight the barriers and attempt to change behaviour through Agenda 21 and other local authority initiatives. The importance of the local as a social concept is central because 'it underpins our attempts to mobilise citizens and their constituencies of interest, together with the networks and partnerships within which they operate at the local level' (Selman, 1996, p. 4). The need for grassroots involvement in local environmental initiatives now seems so 'blindingly obvious as to constitute a truism' (Haughton and Hunter, 1994, p. 303). The challenge for sustainability is to overcome 'our deeply ingrained civic sclerosis and persuading people to modify their actions' (Selman, 1996, p. 158). The next step involves the development of a new form of environmental citizenship which goes 'beyond traditional notions of public-spiritedness' in order to achieve this it is important that 'local citizens have an interest in, even passion for, the "global village"' (Selman, 1996, pp. 15 and 21).

There is a need to strengthen citizenship to foster a sense of 'local community responsibility for urban environmental conditions' (Haughton and Hunter, 1994, p. 224). Local sustainability can 'only be secured if there is widespread mobilisation of people, their energy and assets' (Agyeman and Evans, 1994, p. 16). Environmental citizens are thus expected to change their everyday habits, be responsible consumers, engage in public debate but it is recognised that not all citizens can 'attain this level of competence' (Selman, 1996, p. 155).

Myth 4: The privatisation and liberalisation of resources conflicts with the new localism

While the early environment movement often saw business as the 'villain' in which production driven by consumption led to 'profligate and unsustainable behaviour' the new localism has embraced the concept of partnership with business in the shift to a more sustainable future (Selman, 1996, pp. 127-143). However the new localism still has major problems dealing with privatised utilities, particularly water and energy, who provide resources that have such a central role in environmental policy. Even where companies do show interest in green consumerism this is often rejected as a 'passive' and 'superficial option' because there is only a limited willingness to consume green products (Selman, 1996, p. 148). Privatisation and liberalisation which is often seen as being totally incompatible with sustainability (see Buckingham-Hatfield and Evans, 1996, p. 177). McLaren argues that the 'environmental record of privatisation in the UK has been poor' (1996, p. 172). In order to maximise returns to investors and the treasury has come before the protection of environmental interest, that lobby groups able to weaken environmental commitments and that government regulation still continues to encourage sales of more electricity and construction new roads.

Myth 5: Sustainability can be achieved by restructuring the design and form of localities

The new localism has placed a lot of hope in the ability of landuse planning, building design, conservation and ecology to develop more sustainable localities (Haughton and Hunter, 1994, pp. 80-122; Mitlin and Satterthwaite, 1996; Selman, 1996, p. 109-126). The physicalist notion that localities are the bounded places in relation to which environmental capacities can be mapped and measured supports demands for new techniques to provide quantitative assessments or indicators of local environmental change (see Wackernagel and Rees, 1996) and the development of planning and design tools to reshape the form and location of local development. Attempts have been made to develop new settlement, design and ecological concepts that could help reduce energy demand and trip generation by encouraging alternatives to the car, the introduction of renewable technologies, more energy efficient buildings and greener, more ecologically diverse cities. Debate has focused on the environmental impact of different type of settlement patterns usually focusing on the relative merits of compact versus dispersed cities or new concepts such as concentrated dispersal. There is a tendency here to search for quick, almost deterministic, fixes to re-design cities in ways which reduce environmental impacts.

Myth 6: The new localism requires more environmental data and new modelling techniques

Much of the new localism is characterised by calls for additional modelling and data collection (Jowsey and Kellett, 1996; Nijkamp, 1994). The physical concept of the locality is important in setting a 'framework for measuring, monitoring, and managing environmental resources' (Selman, 1996, p. 4). Research resources in both academic and policy making sectors have been spent on efforts to collect new data and develop new modelling techniques. The new localism is being increasingly characterised by a 'suite of distinctive methods of investigation and analysis' (Selman, 1996, p. 57). Central to this shift is the development of a set of 'decision-support' techniques which provide information about the environment. These include state of the environment reports, quantitative indicators and internal audits of environmental performance. Attempts have been made to develop assessments of the level and extent of the urban carrying capacity or ecological footprint of a locality or city (Wackernagel and Rees, 1996). Such efforts could help produce a new 'sustainability index' (Haughton and Hunter, 1994, p. 233). The main focus is on pulling together disparate information to develop quantitative indicators which map and measure environmental change on a local basis 'with which citizens can more readily identify' (Selman, 1996, p. 1).

Beyond the Myths of the New Localism

The proponents of the new localism are aware of many of the tensions in the approach. Paul Selman acknowledges that many of its assumptions are often 'shrouded in meaningless rhetoric' (1996, p. 157). In a spirit of positive criticism we would now like to briefly sketch out some of our concerns about what is missing from the new localist understanding of environmental processes.

Very little of the 'locality' argument is framed within the context of any wider theoretical framework. Paradoxically the emergence of the new 'localism' contrasts sharply with the social-spatial focus adopted in environmental social theory. Theories of ecological modernisation stress the importance of national and international regulation encouraging a shift towards new forms of production and consumption cycles which minimise environmental impacts (Hajer, 1995; Spaargaren and Mol, 1992), while the risk society perspective stresses the importance of the shifting relations between individuals and global ecological shock (Beck, 1992). Both approaches stress the importance of considering the role of individual, national and global relations and the relations between them, which cannot be easily captured with the new localist agenda. The new localism often reifies the city and by making the assumption that much of the social relations can be captured and contained at local level while ignoring the importance of the national state and globalisation. So, while it might be 'good' to think globally you can safely ignore the wider national and global difficulties and contradictions by acting locally to promote sustainability.

If we are to privilege local authorities as the principal actors in environmental management then it is vital to understand the wider organisational and political context within which local government has adopted the new localism. Agenda 21 and environmentalism has been embraced in the UK in large part in response to the vacuum created by shifts in local governance. During a period of increasing centralisation local authorities were often glad to seize on the new agenda and the role it gave them. Local authorities have then sought to redefine environmental issues for their own purposes - especially in strengthening their importance relative to central government (Selman, 1996, p. 88). The new localism in environmental policy is important because it 'represents an alternative to the Conservatives minimalist vision of local government and gives a new basis of legitimacy' (Ward, 1996, p. 131). Consequently environmental quality has become a key issue in the place marketing of cities through a myriad of new environmental awards. With private sector support the environment becomes a commodity that can be used to differentiate cities. It presents an opportunity for local government to re-invent itself around a new theme in an era when local government has lost powers and control over many aspects of local affairs. Meanwhile, the trend towards reducing local government financial resources and fragmenting and reducing local powers has largely worked against large scale urban environmental strategies (Haughton and Hunter, 1994, p. 301). But in taking up this challenge for its own purposes we must ask the question of whether or not local authorities are actually up to the job. As Evans and Rydin (1997) have pointed out, the '*we know best* implicit in much professionalised planning of the post war period does not sit easily with the post Rio rhetoric of empowerment, capacity building and partnership'.

In the stress on promoting sustainable lifestyles there appears a powerful homogenising ethic and strong sense of social control. Citizens, apparently, need to be forced to adopt a particular lifestyle. Yet individual attitudes, associations and patterns of behaviour are formed in complex ways and their sense of belonging is not necessarily linked to a geographically defined community. Rather individuals relate to a multiplicity of groups which spill over local boundaries and which have the potential to become associations (Ward, 1996, p. 140). There is much more diversity in how the local is constructed by different communities and variation in the reference points or social associations through which life is shaped. The new localism then has an oversimplified view of social change which severely restricts the possibilities for creating new social contexts in which environmental innovation could take place. Despite the protestations of partnership and involvement of new actors there is little evidence of the new localism really attempting to assess what factors actually shape the choices and behaviour of individuals. There is little understanding of the ways in which individuals are embedded in particular types of consumption and production cycles that transcend local boundaries and local institutions. The rhetoric of partnership and new communities then belies a discourse of social control and standardisation. Persuading individuals in isolation is surely not the way forward. Rather than focusing on individual exhortations, for instance in discouraging people from driving to work, there is a need to focus on relevant associations, local employers, chamber of commerce, local unions and employees associations. Working through associations would 'recognise the reality of fragmented identities; avoid parochialism because of the way associations typically extend over local government boundaries or are federated into national and international associations; and potentially more democratic' (Ward, 1996, p. 143).

Although much of the rhetoric of the new localism focuses on partnership with the private sector the practice often fails to open new dialogue with some of the most important and powerful shapers of resource flows in cities that have links to almost every consumer. In particular the new localism often fails to recognise the significance of wider social shifts in which resources are managed. In particular, privatisation of water, sewage and energy services and the increasing role of the private sector in provision and management of transport services has created a new context within which we need to think about environmental policy. Within the localism there is only general, and somewhat grudging, acknowledgement of the potential environmental benefits of these shifts. There is now a need to catch up with the shifts that have taken place in the social organisation of essential resources and develop a new dialogue with utilities that may help identify forms of commercial action that are environmentally beneficial. Here we can think of the ways in which private companies are developing more sophisticated views of their customers and are developing expertise to market energy efficiency services to particular market segments. Such techniques can help develop a markedly different view of the role of customers that goes beyond simple calls for changed behaviour to some new form of sustainable lifestyle that characterises much of the new localism. The private sector could find new ways of making integration work without the need for legislation putting the local authority back into control.

Similarly restructuring the design and form of localities is much more complex than first thought (Blowers, 1993; Breheny, 1992; Voogd, 1994). The relationships between density and energy use are complex and often contradictory and with the slow turnover of urban built form environmental problems cannot simply be re-designed out of cities. Rather than look for physical solutions in the form of simple shifts towards new patterns of urban development a different style of planning is required which creates a social context for energy efficient buildings, alternatives to car based travel and incorporate the environment into a different conception of the planing process. Equally, the efforts of scientists and engineers in modelling the urban environment can be seen as an attempt to reassert a rational scientific approach to policy development as local authorities attempt to physically tame and control the city without much reference to wider social processes. This means that the policy scenarios supported by urban models are often divorced from the changing social organisation of resource flows. Again this approach reasserts the primacy of the locality container model without analysing how the city sits within a wider institutional and social framework. Moreover, the data planners spend months attempting to assemble is often already held by private companies who have considerable data and intelligence on resource use.

Conclusions

We have suggested that in the new localist discourse the 'locality' becomes viewed as some sort of container or black box which can be physically and socially shaped to deliver a more sustainable future through the institution of local government. Rather than examining how the local sits within particular types of national and international relations there is a powerful tendency to develop a much more inward looking approach which divorces the locality from its wider context. By contrast we have argued that the new localism provides only a partial understanding of the local environment and at worst can provide a misleading guide to policy development. Although many of the criticisms we have outlined are to some extent acknowledged in the literature the myths which define the localist discourse have become so prevalent as to represent a conventional orthodoxy. Moreover, as many analysts supporting the new localist philosophy are also promoting the development of a new role for local government these myths may be constricting the scope of local environmental policy. By focusing so narrowly on the role of local authorities and citizens the new localist agenda only confuses the task of constructing sustainable cities.

Simply replacing the 'old structuralism' with a 'new localism' will not do. Instead we urgently need a more critically ambitious debate based on a more theoretically informed and empirical view of the social, political, commercial and technological shaping of local environments. In building local environment policy we need to draw in new participants and ask new questions about what a policy might achieve and for whom. Here we must be sensitive to the ways in which changing commercial discourses and institutional practices may present new opportunities for environmental management. In this way we can work towards a better understanding of how the meaning of the 'local' is constituted by a range of public and private actors in a variety of institutional and commercial contexts at different spatial levels. Such an analytical shift would refocus local environmental policy on the changing social organisation of resources, the restructuring of consumption and production cycles and the relationship between the local, national and global.

Acknowledgements

We would like to thank Paul Selman and Neil Ward for their helpful comments on the ideas developed here.

References

Agyeman, J. and Evans, B. (1994), 'The new environmental agenda', in J. Agyeman and B. Evans (eds), *Local Environmental Policies and Strategies*, Longman, London, pp. 1-22.

Beck, U. (1992), *Risk Society*, Sage, London.

Blowers, A. (ed) (1993), *Planning for a Sustainable Environment*, Earthscan, London.

Breheny, M.J. (ed) (1992), *Sustainable Development and Urban Form*, Pion, London.

Buckingham-Hatfield, S. and Evans, B. (eds) (1996), *Environmental Planning and Sustainability*, Wiley, London.

Evans, B. and Rydin, Y. (1997), 'Planning, professionalism and sustainability', in A. Blowers and B. Evans, *Town Planning in the 21st Century*, Routledge.

Hajer, M.A. (1995), *The Politics of Environmental Discourse*, Oxford University Press, Oxford.

Haughton, G. and Hunter, C. (1994), *Sustainable Cities*, Jessica Kingsley, London.

Jowsey, E. and Kellett, J. (1996), 'Sustainability and methodologies for environmental assessment for cities', in C. Pugh (ed) (1996), *Sustainability, the Environment and Urbanisation*, Earthscan, London, pp. 197-227.

Mitlin, D. and Satterthwaite, D. (1996), 'Sustainable development and cities', in C. Pugh (ed) (1996), *Sustainability, the Environment and Urbanisation*, Earthscan, London, pp. 23-61.

Nijkamp, P. and Perrels, A. (1994), *Sustainable Cities in Europe*, Earthscan, London.

Pugh, C. (Ed) (1996), *Sustainability, the Environment and Urbanisation*, Earthscan, London.

Selman, P. (1996), *Local Sustainability*, Paul Chapman Publishing, London.

Spaaragren, G. and Mol, A.P.J. (1992), 'Sociology, environment and modernity: Ecological modernisation as a theory of social change', *Society and Natural Resources*, vol. 5, pp. 323-344.

Voogd, H. (Ed) (1994), *Issues in Environmental Planning*, Pion, London.

Wackernagel, M. and Rees, W. (1996), *Our Ecological Footprint: Reducing Human Impact on the Earth*, New Society Publishers, Canada.

Ward, H. (1996), 'Green arguments for local democracy', in D. King and G. Stoker (eds), *Rethinking Local Democracy*, Macmillan, London, pp. 130-157.

Webber, P. (1994), 'Environmental strategies', in J. Agyeman and B. Evans (eds), *Local Environmental Policies and Strategies*, Longman, London, pp. 47-61.

19 Local-International Partnerships in Metropolitan Environmental Management: Latin American Experiences

ADRIANA ALLEN

Introduction

In the course of the 1990s international development agencies have become increasingly interested in cooperating directly with local authorities around the themes of environmental planning and management (EPM). However, this interest has so far progressed little in terms of resourcing and facilitating strategies, particularly where the theme of urban sustainability is concerned. Furthermore, few if any large cities in Latin America have dedicated the resources or possessed the institutional vision - as illustrated in the well-publicised case of Curitiba in Brazil - that are necessary to incorporate the environmental dimension effectively into the complex tasks of urban management.

This paper looks at a number of experiences in metropolitan inter-municipal cooperation in Latin America. These constitute of a 'laboratory' in search of integrated approaches to urban EPM which are exploring means to broaden local capacities and exchange experiences with other metropolitan areas. The paper concerns itself with three issues:

- how to understand the implications of environmental management and sustainable development (SD) as it affects rapidly growing and increasingly complex metropolitan areas;
- the specific challenges posed by the concept of sustainability in the context of Latin America; and
- gauging the extent to which urban EPM is moving beyond narrow environmental impact mitigation programmes to develop new

institutional frameworks aiming at the integration of sustainability into the terms of reference of "development" as such.

The Challenge of Sustainability in Latin American Metropolises

First there is a need to present two dimensions of the crisis of sustainability. The first of these, referred to as 'primary sustainability', concerns the sustainable management of the natural resources base. In Latin America the depletion of natural resources is very evident and may be seen to be strongly connected particularly with the future of agricultural production and with the expansion of human settlements.

The second dimension of the sustainability crisis, referred to by some authors as the crisis of 'secondary sustainability' (Neira, 1995; Fernández, 1997) concerns the problem of inadequate management of the 'urban techno-structures' necessary to support human life and productive activities. This secondary crisis is clearly visible in the metropolitan areas with their hyper-concentration of population, highly vulnerable to environmental problems.

The primary and secondary sustainability crises are strongly linked. In the 1950s migration flows from the countryside to urban areas were fuelled by the failure of attempts throughout Latin America to intensify agricultural production and to process the produce through a network of agroproduction services centres. Since the 1960s, the crisis of secondary sustainability has become increasingly evident in the larger cities of the region. This is characterised by a dramatic imbalance between the in-migration flows and the carrying capacity of the cities in terms both of natural resources and techno-structures.

Today in Latin America, three out of four inhabitants live in urban areas (Dávila, 1996) and a high percentage of the urban population of the region is concentrated in large metropolitan areas. Accompanying the massive process of urbanisation that has resulted in the current situation, national economic growth in the region has become increasingly reliant on the urban centres and particularly on the performance of the largest metropolises. These metropolises are subjected to accelerating processes of social, economic, political and environmental change, constantly redefining their conditions of competitiveness, equity, governance and sustainability.

The process of rapid urbanisation experienced in Latin America has been characterised by a weak relationship between the development of urban systems, their hinterlands and ability to take advantage of positive aspects of modernisation. In this context local governments are facing a new situation. The development process, particularly in metropolitan areas, seems to be increasingly subordinated to national macroeconomic policies and the decisions and movements of transnational capital. At the same time increasing demands for democratisation and decentralisation have pushed local governments to abandon their traditional role as urban administrators and to assume the challenge of improving the environment and social conditions of their citizens by attempting to enhance the quality of local governance.

Although metropolitan areas pose some of the most critical challenges, it is important not to forget the importance of intermediate cities (between 500,000 and one million). These cities are growing at faster rates than those over one million and in most cases their growth is not accompanied by an effective decentralisation of regional and national productive activities. They possess an inadequate industrial base, having to rely on footloose inward investment where they are generally at a disadvantage relative to the larger metropolises. In addition, these cities are severely lacking in financial resources with municipalities spending only a tiny fraction of the national budget.

Urban authorities are under considerable pressure from various, sometimes conflicting, interests to improve their performance. This translates into attempts to reform the administrative process with an orientation not only to solve present problems but also to look towards solving longer term problems of an integrated approach to sustainable development.

Inter-municipal Cooperation: An Emerging Approach in Urban EPM

The role played by national and international associations of municipalities has been key in fostering inter-municipal - or decentralised - cooperation. In the last ten years this activity has increased dramatically. This might be ascribed to the pressures posed on the one hand by social, economic and environmental 'globalisation' and on the other by the increasing responsibilities assumed by the municipalities through processes of decentralisation. Increasingly, networks of cities and towns around the world are being integrated into a global structure of international relations.

The process of inter-municipal cooperation in EPM has been developing via two mechanisms. On one hand, urban local authorities sharing particular problems or interests have established 'twinning' or 'jumelage' arrangements to promote bilateral exchanges of experiences, technical assistance or other mutual activities. On the other hand, many donor agencies and international associations of municipalities have developed new programmes aiming at establishing networks of cooperation amongst different cities with a common interest inter alia of fostering sustainable development.

In Latin America, decentralised cooperation programmes are involving several UN agencies and European bilateral development agencies (Bidus, 1995). In 1994 a two week mission held by the staff of the International Council for Local Environmental Initiatives (ICLEI, 1995) reviewed some of these experiences and concluded that Latin American cities are at the forefront of initiatives inter-municipal cooperation around sustainable development. Nevertheless, only relatively few countries in the region have adopted this approach and even within these countries these new aproaches have only involved a few cities.

The UN Conference on Environment and Development (UNCED - 1992), and in particular Agenda 21 signed by most countries attending the conference, has been an important impetus to these programmes. The importance given by Agenda 21 to local action, which is being promoted in the form of 'Local Agenda 21' (LA21), has been an inspiration to many new and innovative programmes promoted by UN agencies and international municipal associations alike.

Looking at the origin of the regional experiences fostering LA21, in most cases they rely on the innovative leadership capacity and international connections of local authorities. In a few cases national authorities have been involved, funding and supporting local programmes from the outset within a national-level decentralisation strategy. In others even when the initiative started at the local level, the national authorities became involved early in the programme in order to address sensitive political issues of municipal powers and decentralisation.

The pioneer local initiatives promoting sustainable urban development in Latin America were the metropolitan cities of Bogota and Curitiba. Both these cities initiated multi-partnership, cross-sectoral programmes aimed at sustainable urban development well before UNCED. In the case of Bogota the so-called Misión Siglo XXI was launched in 1990 to promote SD in the 20 districts comprising metropolitan Bogota. This experience was institutionalised in 1994 via the establishment of a non-profit organisation in order to promote and assist in the implementation of a LA21 process in other Colombian municipalities. This activity was coordinated with a radical national decentralisation programme.

Prospects to Encourage Intermunicipal Cooperation towards SD

The emergence of municipal networks across international boundaries focused on promoting urban sustainability is a new phenomenon; it is therefore too early to provide any overall assessment of their strengths and weaknesses. However, the experiences reviewed provide some ground to reflect on the contribution of this mechanism to addressing the environmental challenges faced by Latin American cities.

Early initiatives in inter-municipal cooperation have greatly promoted the implementation of decentralisation programmes, new multi-partnerships and a cross-sectoral approach reshaping the role of local authorities and fostering new local management abilities. These three components are particularly relevant to improving metropolitan EPM. From an institutional point of view metropolitan areas generally include overlapping jurisdictions with weak links and limited municipal power in key areas such as transport, water supply, energy, solid and liquid waste management and land use planning.

For instance Rio de Janeiro metropolitan region consists of 13 municipalities, Buenos Aires has 20 and Mexico City is administered as 27 municipalities and 16 delegated areas of the Federal District. In some cases these municipalities are managed by separate higher-level administrative units. Conflicts between national, provincial and local authorities is common. In Buenos Aires for example the provision of basic urban services and infrastructure is split between various national government agencies and the Province of Buenos Aires (Gilbert, 1996).

In social and economic terms the various service sectors are integrated through complementary industrial and service functions between neighbouring districts and increasing numbers of commuters. Therefore the initiatives attempt to acknowledge and work with metropolitan management complexity beyond narrow jurisdictional and political boundaries.

The lessons of these experiences are also meaningful for other urban areas. In the majority of the cases urban EPM demands a cross-jurisdictional and cross-sectoral approach. Many intermediate cities in Latin America will become metropolitan areas in a few decades and already rely on the decisions taken in neighbouring municipalities to guarantee, for example, fresh water and food for their inhabitants. Therefore, the "metropolitan" status should be seen not only as a function of high concentration of population in a continuously urbanised area but also as a system of regional functions linking smaller cities.

The emergence of the so-called "regional metropolitan areas" in Latin America asserts the need to rethink the traditional definition of metropolitan areas. The case of the Upper Valley Metropolitan Region in the North of Patagonia in Argentina clearly illustrates a settlement pattern where metropolitan functions for a vast region are performed by a group of three intermediate size cities and several small towns (Vapnarsky and Pantelides, 1987).

Regional metropolitan areas are characterised by dispersed and yet highly integrated settlement patterns within which several towns and cities with relatively low concentrations of population perform complementary functions in the provision of services and infrastructure and in the development of economic activities. Metropolitan EPM demands a conceptual and methodological approach to move away from the physical definition of cities to a broader understanding of the articulation of complex patterns of settlements, where the flow of natural resources, capital, goods, services and people is not restricted by jurisdictional boundaries.

Most of the Latin American municipalities undertaking LA21 initiatives are placing emphasis upon the creation of new institutional frameworks and multi-sectoral participatory partnerships. These are essential prerequisites to foster the creation of a permanent institutional capacity to implement long-term EPM processes towards sustainable urban development. Intermunicipal cooperation has proved to be a powerful way to break vertical national-local hierarchies and competitive approaches between municipalities. Networking with international partners allows a better dialogue between national and local governments and between local authorities, facilitating the exchange of resources, methodologies and technical assistance.

In addition it empowers local authorities to establish direct relations with international donor agencies overcoming powerless in the past to obtain external funds. The international donor community is reshaping its strategies and starting to orient specific urban development funds to support initiatives defined according to local priorities. However, the lack of articulation between local and national policies makes one wonder whether these initiatives are genuinely contributing towards SD or just enhancing local management skills without addressing the key challenges posed by the concept of sustainability in the world of economic and cultural globalisation.

As argued above, urban sustainability cannot be addressed without considering regional and national development trends and the extent to which the management of natural resources is subordinated to economic globalisation. When a given volume of natural and built resources is insufficient to sustain urban populations with a given pattern of production and consumption there are three possibilities to rectify this: through territorial expansion, technological innovation or the restructuring of unsustainable production and consumption patterns. So far, the first two alternatives have been seen as the answers. These do not necessarily confront the urban sustainability crisis however, but displace its effects across time and space. According to Fernández (1997) the third approach poses several challenges partially addressed by the experiences reviewed as follows:

- The challenge of ecological sustainability, understood as the rational management of natural   resources use, and of the pressures exerted by the wastes produced by society, which demands an integrated view of local, regional, national and international trends in environment and development.

- The challenge of social sustainability defined as a set of actions and policies oriented to the improvement of social quality of life, but also to the fair access and distribution of rights over the appropriation and use of natural and built resources.

- The challenge of political sustainability, characterised at the micro level as the democratisation of the society and at the macro level as the democratisation of the state.

It is difficult to draw the line between initiatives focused on a traditional approach to urban EPM and those which are genuinely working to deliver sustainability. If we apply the three aforementioned criteria to the ongoing LA21 initiatives in Latin America the picture reveals more challenges to be addressed than as yet showing any positive achievements

Some of the local efforts under way reveal that local governments have now better understood that LA21 implies a very different approach from more traditional local environmental policy formulation and implementation, and there are a number of front-running municipalities which appear to be pursuing SD goals which are more far-reaching than those of their national governments. However, a key factor to be addressed - in which inter-municipal networking could play a major role - is the development of increasing efforts to put pressure on national governments to work seriously towards the realisation of SD goals and to further enhance the implementation of the UNCED agreements. Without the active engagement of national authorities in support of local initiatives, the political legitimacy and viability of LA21 might be threatened in the long run, diminishing the further development of local sustainability strategies.

The strength of national government support for LA21 is a vital factor influencing the success of local authorities in promoting and developing LA21 processes, and to articulate local initiatives and national policies towards SD. However, in the Latin American experience the establishment of national committees to promote and coordinate LA21 implementation and the allocation of special national funds to this end are still rare. Another role for national governments consists not only in supporting LA21 initiatives in financial and political terms - through decentralisation of power and money - but also in building up local capacities and enabling and promoting the exchange and dissemination of local practices.

A second challenge is the small amount of attention that has been paid so far to the use of inter-municipal cooperation mechanisms among small and intermediate cities. This could arrest the intense concentration of urban life in a few megacities promoting a more sustainable and diversified pattern of settlements. Even successful cases of metropolitan EPM usually rely on the improvement of local environmental conditions but without addressing the reduction of their wider resource and environmental impacts.

Concerning the third challenge, most LA21 initiatives in Latin America are focused on preserving local settings or restoring water and air quality. This remains far from any attempt at deep-seated changes in current production and consumption patterns. Very often metropolises protect their own local environments through their ability to appropriate (or purchase) additional ecological capacity from somewhere else. Metropolitan EPM must address the challenge of improving local conditions without eroding the ecological capacities beyond their boundaries.

But such a shift in the development and management paradigm requires the use of new analytical tools, such as the concept of 'ecological footprint' (Rees, 1992) or 'environmental space' to provide decision-makers, managers and the general public with an insight into the magnitude of the current ecological impact and vulnerability of metropolitan areas. These must also point to available opportunities for reducing the footprint of urban activities and lifestyles through intelligent urban development and management.

So far, the LA21 movement in Latin America (and probably elsewhere) has not fully addressed the challenge of sustainability. Most experiences have focused on strengthening governance, fostering the necessary conditions to develop new institutional frameworks which provide innovative ways to address urban EPM in a time of "municipalisation" of the crisis. Participation, partnership and networking appear to be the cornerstones of a new management paradigm. Although new ways of understanding and managing urban and metropolitan systems are emerging within the LA21 movement, most experiences are still short of policy aspirations to address the challenge of primary and secondary sustainability. In other words new tools and approaches for urban EPM are rapidly being introduced but no shift in the development paradigm can yet be discerned.

References

Bidus, M. (1995), 'A review of decentralised and municipal development initiatives and their effect on democratisation in Central America', Washington DC, United States Agency for International Development.

Dávila, J. (1996), 'Enlightened cities: The urban environment in Latin America', in Hellen Collinson (ed.), *Green Guerillas*, Latin America Bureau, London.

Gilbert, A. (ed.) (1996), *The Megacity in Latin America*, United Nations University Press, New York.

Gilbert, R., D. Stevenson, H. Giradet and R. Stren (1996), *Making Cities Work. The Role of Local Authorities in the Urban Environment*, Earthscan, London.

International Council for Local Environmental Initiatives (ICLEI) (1995), *The Role of Local Authorities in Sustainable Development: 14 Case Studies on the Local Agenda 21 Process*, UNCHS, Nairobi.

Neira, E. (1995), *La Sustentabilidad de las Metrópolis Latinoamericanas*, PNUMA - El Colegio de Mexico, Mexico.

Rees, W. (1992), 'Ecological footprints and appropriated carrying capacity: What urban economics leaves out', *Environment and Urbanization*, vol. 4, no. 2 (October).

United Nations Centre for Human Settlements (UNCHS) (1994), *Sustainable Cities: Concepts and Applications of a United Nations Programme*, UNCHS, Nairobi.

Vapnarsky, C. and E. Pantalides (1987), *La Formación de un Area Metropolitana en la Patagonia. Población y Asentamiento en el Alto Valle*, Centro de Estudios Regionales, Buenos Aires.

20 The Local Agenda 21 Process in the United Kingdom: Lessons for Policy and Practice in Stakeholder Participation

CHRIS CHURCH

The Local Agenda 21 Process in the United Kingdom

Since at least the mid 1980s, increasing attention has been paid to the need to rethink and reorganise the process of planning and management so as to take into consideration the problematic of 'sustainable development'. Defined by the World Commission on Environment and Development as 'development that satisfies the needs of the present generation without jeopardising the needs of future generations', this has often been construed simply as a call for improvement in environmental management.

However, it is increasingly recognised that to satisfy the basic principle of sustainable development it will be necessary to revise more than just the management of the environment: it will require radical revision of lifestyles, production and the use of resources in general.

The UN Conference on Environment and Development in 1992 made this point in a more concrete way in the pages of Agenda 21, the 600 page 'agenda for the 21st century', signed by the governments which attended the conference. Agenda 21 took the theory of sustainable development and attempted to turn it into a workable action plan. Any serious reading of this document makes it clear that if the theory is to be turned into action, much of this will have to take place at the local or grassroots level. With that in mind Chapter 28 of Agenda 21 states that 'by 1996 most local authorities in each country should have undertaken a consultative process with their populations and achieved a consensus on "a local Agenda 21" for the community.' The call to engage in Local Agenda 21 - or simply LA21 - processes on the part of local authorities and communities has, arguably, been the most successful line of implementation of Agenda 21. However, the call for such processes to be undertaken brings together three policy areas in which the key concepts are all contested – community, sustainable development, and participation. Each of these words and the ideas behind them are the focus of debate: it is therefore unsurprising that the development of Local Agenda 21 plans has been a process with many different approaches and visions.

Has it Been a Success?

The Local Agenda 21 process has now been running in the UK for over five years. Across the country substantial resources have gone into it, notably local council staff time and money, resources of organisations within civil society and time donated by individual volunteers. The need for a clear analysis is now urgent. Yet any assessment of the success of Local Agenda 21 in the UK must ask what is meant by "success"?

From the beginning there was a lack of clarity in the purpose of LA21. This is not surprising given that it is a fundamentally new approach to local development with no established procedures. The original aim, as expressed in Agenda 21, was a 'Local Plan for Sustainable Development' that would focus on the key issues in that document, including poverty, health, and livelihoods as well as resource and environmental issues.

However the idea of LA21 came into use at a time when many UK local councils were for the first time considering their broader environmental responsibilities, and many were at work producing Environmental Charters, Action Plans and so on. Local Agenda 21 was seen as a direct development out of this, and most LA21 work was, and remains, in the hands of local authority "Environmental Co-ordinators". This has led to the criticism from workers in the public health and other relevant sectors that in some areas Local Agenda 21 has been 'hijacked by environmentalists'.

Two central questions for any evaluation must be:

- How far is this environmental focus drawing attention away from the socio-economic issues that Agenda 21 identifies as central to sustainability?
- How far is it possible for one process to shoulder all the problems and responsibilities covered by Agenda 21?

Who is Doing LA21?

Basic statistics have been collected by the University of Westminster for a Local Government Management Board (LGMB) funded survey (Tuxworth, 1996) concerning the extent and content of Local Agenda 21 initiatives in the UK. The figures from surveys in February and November 1996 suggest that 347 authorities out of 475 (73%) have made some commitment to Local Agenda 21 programmes. However this commitment varies enormously: only some 140 appear to have appointed a staff officer with specific responsibility for LA21. Twenty-four authorities claim to be integrating sustainability principles into internal working, although about 160 claimed to be developing sustainability indicators. Most local authorities accepted that to date Local Agenda 21 has had a small, very small, or no impact on key issues such as health, meeting basic needs, or social services.

Inevitably such statistics only tell a small part of the story. A dispassionate overview of the surveys suggests that around 33% of responding councils – about 100 in total - are working actively with their communities and are attempting to work on sustainable development rather than purely environmental concerns.

While this may not be the full 73% of local authorities involved in some way with LA21, it nevertheless implies a very significant body of activity – perhaps the largest network in any country of processes that are identifying themselves primarily as locally-managed LA21 programmes. This apparent significance indicates the need for a clear assessment.

Work in Progress

While there is no set framework, material produced by various bodies, including the International Council for Local Environmental Initiatives (ICLEI 1996), Chartered Institute of Environmental Health (CIEH 1995), United Nations Association-UK's *Sustainable Development Unit* (Church, 1995) and most notably work by the Local Government Management Board (LGMB), has suggested ways forward that have led to certain common approaches being widely adopted. The key components of these approaches include:

- *consultation with interested groups:* this is the commonest approach. Councils set up working groups, usually on specific themes such as Energy, Transport etc. These identify key issues and come up with recommendations for action. Such groups have tended to attract professionals and NGOs with a defined interest; they have not often been an effective way into involving a broader public;

- *consultation with the wider public:* how far the public is involved varies widely. Some councils have set up innovative exercises, while others have done nothing more than circulate leaflets. Several participation tools, such as *Planning for Real* (RIBA, 1997), have been used in this context, with widely varying degrees of success. Very few councils have clear policy guidelines in this area.

- *internal environmental management by local councils:* it is accepted that if local councils are to promote lifestyle and policy changes, then they must be seen to be active themselves. Most (85% of respondents in a recent survey by researchers at Westminster University) claim to have some form of systematic approach to this work, although only 12% (about 35 councils) are registered with the heavily promoted European Community inspired Environmental Management and Audit Scheme (EMAS) system;

- *production of a 'Local Agenda 21' document:* production of some form of plan or document is an important part of the LA21 process. Such documents set out the objectives and goals of the process, and show those who have given time and energy that there is a clear output. Over 100 Local Agenda 21 documents now exist and many more are already available as drafts or reports; and

- *the use of indicators:* in 1994-95 substantial research went into the generation of local indicators of sustainable development involving a number of local authorities, supported by the LGMB, along with the UNA SDU and the New Economics Foundation. This stimulated interest in indicators more generally (Cartwright, 1997). However, it is not clear how far the use of such indicators is being taken forward by councils as a whole, or being linked to policy decisions.

The Viewpoint of the Central Government

Stakeholders Speak Out – The UK Citizen's Report

One consideration with regard to the assessment of the success or otherwise of Local Agenda 21 must be the results of capacity-building processes. Many community-based and non-governmental organisations involved in developing Local Agenda 21 initiatives are now developing their own priorities. In May 1997 UNED UK undertook a consultative process in the run-up to 'Rio Plus 5', the follow-up conference to UNCED. Some fifty local networks responded with their own priorities for actions that Governments and the UN should take to support Local Agenda 21 activities in the next five years.

The issues raised by these organisations and presented to the UN in the 'UK Citizens' Report' (UNED-UK, 1997a) show a rather different agenda to current environment-led activity. The main points were, in order of priority:

- citizenship and participation;
- poverty and exclusion;
- transport;
- energy;
- sustainable consumption and production;
- global development, aid and trade; and
- valuing the whole community.

It is clear that participation and partnerships - the need to open up the local development process to may different actors - is a widespread concern. The best experiences in Local Agenda 21 have, indeed, involved different stakeholder groups as equal partners with local authorities, countering those critics who see local government as remote and unresponsive.

However, many LA21 networks admit that they have as yet failed to build an alliance that represents a genuine cross-section of the whole community. Specific areas of concern are the under-involvement of black and ethnic minority communities, poorer communities, youth and the aged. The non-involvement of such groups is a major failing of participative processes that have developed with little forward planning or policy (Taylor,

1995). It is important to realise that participatory approaches to LA21 are dependent upon the degree to which elected officials are prepared to relinquish their powers to other groups in the community.

Evaluating Local Agenda 21: The Search for Success and Good Practice

The Need for Success Criteria

Local Agenda 21 is by no means the panacea that some of its advocates claim it to be. In many parts of the country it is invisible to the public, often even in those areas seen by "experts" as examples of good practice. The idea that all political decisions are best taken at a local level (a common "localisation myth" discussed elsewhere in this book by Marvin and Guy) is clearly an exaggeration.

Nevertheless, much work has been done, and there is a need for a commonly agreed framework in which to evaluate the effectiveness of this. In the first instance, whilst Sustainable Development might be the ultimate goal of Local Agenda 21 processes, environmental improvements should be the first evidence to emerge. If there is no evidence of such improvements then the value of LA21 must be questioned.

However, it is also important to remember that LA21 is defined as a "process of consultation" between local authorities and their communities. Any assessment should therefore consider the nature and quality of consultation, the ideas and proposals that have resulted from it, and the ways in which these have brought about genuine lasting change.

The search for good practice is going on in many areas. Given the absence of any statutory duty on local councils to carry out this work, showing how good practice elsewhere has improved the localities where they have emerged is one of the few means to persuade lagging councils to become involved. Yet "good" or even "best" practice is not always easy to define. The LGMB's Case Studies project (LGMB, 1997) relies on getting majority agreement from ten NGOs which act as advisers, but there are as yet no formal criteria for determining what is or is not a "good practice". Other bodies have adopted an attitude of: 'this is best practice because we say it is' or have developed "sustainability checklists" that apply to projects rather than to the Local Agenda 21 process as a whole.

The Questions to be Asked

Success in LA21 processes should consider three important areas and ask hard questions regarding:

- *Process:* Has the process of consultation been designed so as to ensure that all stakeholders had a genuine opportunity to take part and have an input?

- *Projects:* Are things actually happening in the locality as a result of the Local Agenda 21 process?

- *Policies:* Are the policies of local authorities and other affected bodies changing as a result of the LA21 process in ways that support moves towards sustainable development?

The purpose here is not to assess the state of Local Agenda 21 initiatives, so much as to set out criteria by which an assessment can be made within particular local areas. Accordingly the fourteen questions below attempt to set out ways in which the three key issues might be considered in an objective manner.

Process

LA21 has opened a great many people's eyes to the problems associated with involving people in discussion regarding positive change. In the great majority of cases LA21 processes in UK local authority areas have failed to significantly involve the broader community, tending to focus on involving groups (principally those concerned with environmental issues) who already have a clear agenda relating to the issues under discussion. People from deprived areas have been conspicuous by their absence, as have representatives of black and ethnic minority communities, despite the evidence that poorer communities suffer disproportionately from environmental problems.

There has also been a very noticeable failure to learn from community development and public health work, where many of the issues on the sustainability agenda have been the focus of attention for many years (Taylor, 1995 and SCDC, 1997). Most "participative" processes have in fact been largely consultative, and are often run by staff who are, furthermore, untrained and under-resourced even for organising effective public consultation.

Good public participation in the context of a Local Agenda 21 process (RIBA, 1997) should perhaps be able to give positive answers in the following areas:

- Is there a clear strategy for public involvement, with a budget and resources available for implementation?

- Has there been a clear assessment of the networks and organisations in the locality that are most genuinely representative of the majority of stakeholders?

- Have the staff responsible for this work been given training where necessary on public involvement?

- Have efforts been made to involve groups who are traditionally under-involved?

- Has the number of organisations actively participating in the process increased with time, and are those organisations having an effective say in managing the process?

- Has any survey been done of how far those people and community-based organisations who have been involved feel that the process has been worthwhile and has led to lasting change?

Projects

If people give their time to a process then they expect to see results. If Local Agenda 21 is to have an impact then it should be leading to projects and programmes that would otherwise not have happened. These will clearly focus on environmental action but might also cover projects linking environmental and social action.

- Are there projects (planned or in operation) that have been developed as a result of the local LA21 process?

- Is there an environmental management programme in operation, with targets and monitoring mechanisms, within the local council?

- Are there projects (planned or in operation) that specifically reflect the cross-sectoral nature of sustainable development?

Policies

Local Agenda 21 must promote and enable environmental change, and a good LA21 strategy should set clear targets for both a local authority and the community it serves. There is little point in the UK Government setting targets, be they for global warming, recycling, or increased use of public transport, if those targets are not reflected in local implementation strategies.

It is also fair to say that if Local Agenda 21 is to help deliver sustainable development then it must not merely cover environmental issues, but should include policies that support action on social equity and economic development issues.

If we are to consider "good practice" in this area then there are two obvious options. One is to say that such good practice entails the adoption of policies and targets, with strategies to achieve them. However it is also quite legitimate to suggest that LA21 targets should at the very least meet and support national targets. A more rigorous approach is therefore to agree that if such targets are to be seen as genuinely 'good' practice (rather than "average" practice) then they should be more ambitious than national baselines.

Such targets will be of little value unless there is a clear strategy for achieving them. This should be identified, either in the LA21 document or in some publicly available support document.

Both these approaches are, for the moment, ambitious. Snapshot surveys suggest that few LA21 processes are including such policy targets, and if they are, then they are not necessarily being adopted by the local authority itself. The questions to be answered include:

- Has the Local Agenda 21 process produced a set of goals and targets with clear specific measurable objectives linked to these targets?
- Have any or all of these targets been adopted as council policy?
- Do the targets on global warming and recycling match or exceed national targets?

- An anti-Poverty strategy is an "integral part" of sustainable development (Agenda 21 Chapter 3). Does the Local Agenda 21 reflect the targets of any local anti-poverty strategy or plan?

- If the LA21 programme has developed a set of indicators are these linked to agreed targets, and are there funded programmes to help meet those targets?

Conclusion

Local Agenda 21 programmes in the UK provide valuable experience as to how the idea of sustainable development can be translated into effective local action. How far there is genuinely effective local action, and how that effectiveness may be judged, is as yet far from clear. Most assessment has merely focused on listing and counting what has been done by local authorities, rather than what has been achieved in terms of concrete objectives.

There is still a strong environmental focus to most Local Agenda 21 activity. This draws attention away from socio-economic issues, but the balance is shifting. It is also pertinent to consider how far Local Agenda 21 is in fact the best way to bring about environmental change. If that was the priority then resources might have been better spent on short, directed programmes to meet easily achievable targets in areas such as energy and resource use. However the long-term goals of sustainability have been better served by developing this wider process.

How far this wider process can cover all the issues in Agenda 21 is still far from clear. If that is to happen considerably more work will need to be done to involve professionals and local user groups in health, poverty, and economic development programmes, whose influence and resources are currently having a more significant impact on societal change at a local level (UNED, 1997b).

The original aim of the Local Agenda 21 process was to produce plans 'by 1996'. It is now clear that production of such plans is only the first step in implementing a programme of change and that such a programme will necessarily be a long one. It therefore makes sense that there should be a full and open discussion about the nature of good practice in this field and how capacity may be built within local councils, NGOs and community-based organisations to enable them to develop communities that contribute to sustainability.

References

Church, C. (1995), *Towards Local Sustainability. A Review of Local Agenda 21 in the UK*, United Nations Association, London.

Abbot, J. (1996), *Sharing the City*, Earthscan, London.

Cartwright, L. (1997), 'The implementation of sustainable development by local authorities in the South East of England', *Planning Practice & Research*, vol. 12, no. 4.

CIEH (1995), *Environmental health for Sustainable Development*, Chartered Institute of Environmental Health, London.

ICLEI (1996), *The Local Agenda 21 Planning Guide*, International Council for Local Environmental Initiatives, Toronto.

Local Government Management Board (1997), *Case Studies for local sustainability*, LGMB, London.

RIBA Community Architecture Group (1997), *Effective Practice in Public Participation*, Royal Institute of British Architects, London.

Scottish Community Development Centre (1997), 'Measuring Community Development in Northern Ireland', Department of Health and Social Security, Belfast.

Taylor, M. (1995), *Unleashing the potential: bringing residents to the centre of regeneration*, Joseph Rowntree Foundation, London.

Tuxworth, B. (1996), 'From Environment to Sustainability: Surveys and Analysis of Local Agenda 21 Process Development in UK Local Authorities', *Local Environment*, vol.1, no.3, pp.277-297.

UNED-UK (1997a), *Keeping Our Side of the Bargain. The UK Citizens Report to the UN*, United Nations Environment and Development committee for the UK, London (available on http://www.oneworld.org/uned_uk)

UNED-UK (1997b), *Building Our Future – Sustainable regeneration in North & East London*, United Nations Environment and Development Committee for the UK, London.

UNEP (1992), *Agenda 21*, Regency Press, London.

21 Negotiating the Local-Central Government Relationship: Experience from Flood Hazard Management[1]

JOHN HANDMER

Introduction

Metropolitan or central governments often develop policies for hazard and environmental management, but depend on local jurisdictions for implementation. The priorities and values of the different levels of government may not match, with the result that important policy initiatives may be ignored or cause conflict. This paper examines two approaches available to higher level governments for policy design and implementation: coercion and cooperation. The cooperative approach is characterised by negotiation and capacity building which takes account of local circumstances. Where local commitment to hazard management already exists, a cooperative approach leads to results equal or better than those achieved under coercion; and cooperative policies may be superior in maintaining local government commitment. Coercion appears to be more effective where local authorities view flooding as unimportant, but its use is often problematic. The paper draws on the results of an international research project which examined urban flood hazard management in jurisdictions with contrasting policy styles.

Flood Hazards and Policy Implementation

Hazard management is increasingly a process where trade-offs between the often conflicting objectives of hazard reduction, economic use and environmental amenity of a hazardous area are negotiated explicitly. The implicit admission is that we are "managing" rather than "preventing" flood hazards. The emphasis in this approach is on accommodating multiple social goals - and on meeting the imperative of local accountability. This requires different skills: those of the negotiator rather than builder are increasingly called for. Many of the newer approaches including that of institutional reform, would not normally be thought of as part of flood hazard management. This highlights the importance of broadening the definition of flood hazard management measures, and flood related organisations. Results of broader thinking include building local capacity for and commitment to land-use measures, modifying the legal liability environment, managing community vulnerability, and involving lending institutions and insurers.

Policies or guidelines for the management of flood-prone land, whether based on concrete, land-use or institutional change, are often developed by higher level governments to satisfy national goals and to counter local recalcitrance. But the various tasks within hazard management may be assigned to different levels of government, so that while the higher level may produce a national strategy, the task of local government is often to implement this strategy. The local implementors can do this in a way that tries to address local concerns as well as higher level priorities, or they can act merely as agents for central government. Typically, local entities possess the necessary planning and land use control powers - although in reality power is generally shared and exercised concurrently by different levels of government, rather than exercised exclusively by one tier or the other (see e.g. May and Williams, 1986; Brouwer, 1994; Healey, 1994).

In summary, gaining local compliance with higher level policies is a major issue in many areas of government, not only within hazard management - and usually involves much more than simply specifying a set of standards or legally binding regulations.

This challenge is examined here in two ways: first, what sort of approaches to *policy design* are available to higher level government seeking to persuade local authorities to implement policies such as land use management; and second, what *attributes* does the lower level government need to help ensure that policies are implemented? The paper draws on aspects of a major international project on the subject reported in full in May et al (1996; also see Acknowledgements). Here, specific examples are drawn primarily from the Australian state of New South Wales (NSW).

Policy Design and Attributes for Implementation

At its most basic, a policy statement comprises two elements: the intentions of the policy - its substantive elements; and the instruments or means through which the intentions are to be implemented - its procedural elements. The focus here is on the means or procedural aspects.

Our international study conceptualised and examined two general approaches to policy design, which differ in the way they work to achieve local implementation (May and Handmer, 1992; May et al, 1996). One approach is based on *cooperation* between the two levels of government. In the other approach the senior level of government employs *coercion* to induce local government to implement its policies. The means or efforts used to obtain compliance with higher level government goals are also dichotomised into approaches labelled the "carrot or stick", "incentives or sanctions", or "persuasion or punishment". This is the case in the context of both private sector regulation, intergovernmental relations, and in areas where the aim of social control is more obvious, such as the criminal justice system (see Bollens, 1993.)

Whichever approach is taken - and frequently the approach in practice is some mixture of the coercive/cooperative dichotomy outlined above - local government as the implementing authority needs certain attributes. These can be seen most simply as:

- the local authority must want to do it, that is they must have *commitment* to the policy objectives.
- they must have the ability or *capacity* to implement the objectives; and
- cutting across both these attributes, there should be a *process* to deal with conflicts between the different interest groups, in particular the conflict between the imperatives of hazard management, economic and social development, and environmental amenity.

At issue is which approach to policy design is most likely to lead to successful local implementation - including development of the necessary local attributes.

Policy Design: Cooperation or Coercion?

Cooperation

The cooperative approach to policy design attempts to make local government a partner in the achievement of higher level government goals (Table 22.1). The emphasis is on enhancing local ability or capacity to reach these goals. Implicit are the assumptions that the local entity is committed to the same ends, and that they will cooperate with higher level government. Emphasis is placed on regulatory or performance goals (the goals might be a specific level of public safety and decreased flood damage potential), rather than prescribed standards (such as the prohibition of all floodplain development or detailed specification of building requirements), under the presumption that local governments will devise the best means within their communities for reaching the performance goals. The approach places the main responsibility for ensuring sound floodplain management on local governments. Table 21.2 summarises some aspects of the current cooperative (since 1984) and previous coercive (between 1977 and 1984) policies in NSW.

Table 21.1   Coercive and cooperative policy designs

	Coercive Policy	**Cooperative Policy**
Policy Objective	Adherence to prescribed standards	Achievement of policy goals
Role of local government	Regulatory agents: enforce rules or regulations prescribed by higher level governments	Regulatory partners (or trustees): develop and apply rules that are consistent with higher level goals
Emphasis of higher level government policy	Prescribe regulatory actions and plans: along with a required process	Prescribe process and goals: specify planning considerations, along with performance standards
Control of local government	Monitoring for procedural compliance: enforcement and penalties for failing to	Monitoring for substantive compliance: financial inducements to develop

	Coercive Policy	**Cooperative Policy**
	meet deadlines or for not adhering to the prescribed process	plans, advice, no penalties
Assumptions about implementation	Commitment of local government is a potential problem; need for uniform application of policies	Commitment is not a problem; local discretion is important in implementation
Implementation emphasis	Adherence to detailed policy prescriptions and regulatory standards. Building "calculated" commitment	Building capacity of local government to reach policy goals. Enhancing "normative" commitment
Potential problems	Weak monitoring of performance and unwillingness to use penalties	Gaps in local commitment and insufficient resources to build capacity. Possibility of "capture".

Source: Adapted from May and Handmer (1992) and May et al (1996, p. 4).

Typically, incentives are offered by higher levels of government for cooperation by lower levels - in contrast to the penalties used in a coercive policy design. Among other things, incentives may be money, technical assistance, or changes to the legal environment. As one inducement to comply with the NSW policy, immunity from legal liability is provided for actions taken substantially in accordance with state policy.

A cooperative approach is inherently flexible, and recognises that the achievement of the goal of flood hazard management involves tradeoffs with other legitimate goals such as economic development, and must have the cooperation of local government. It has the ability to retreat in the face of strong opposition, while remaining ready to make progress as opportunities arise. Inevitably, this orientation requires increased use of negotiation and conflict resolution skills, which themselves require a guiding framework. The NSW policy sets out a process for the development of a local floodplain management plan. This provides the necessary framework for negotiation primarily through committees of stakeholders, and the preparation and circulation of reports for public comment. In addition, environmental impact statements are required for major engineering works.

Potential difficulties with this approach are the need for adequate funds and expertise to build implementation capacity; the possibility that the flexibility may result in a gradual rise in the damage potential due to the cumulative results of many minor "exceptions"; the possibility that development interests might "capture" the process; and the lack of penalties to use against recalcitrant local governments who do not cooperate. Another potential difficulty is that ecological "bottom lines" might be traded against economic development or some other priority - although this is a characteristic of most policy anyway.

"Capture" by commercial development interests, whereby the processes are manipulated for commercial gain, did not appear to be widespread in NSW - although case studies indicate that it may be important locally. It appears that negotiating solutions is not resulting in substantial increases in flood damage potential - at least within the 1 percent (or 100 year) floodplain. Non-complying local governments in areas of rapid urbanisation and therefore potentially serious new flood hazards, pose a dilemma for those who want to avoid coercive strategies. One approach is to establish and strongly support local committees designed to bring floodplain stakeholders together. These committees can then start developing a flood hazard management plan. Evidence suggests that the chances of success are enhanced when such cooperative strategies are used by central government. Conversely, inflexible legalistic approaches in these circumstances were found to discourage local cooperative effort.

Table 22.2 NSW flood policy design, pre- and post-1984

	1977-1984	Post 1984
Regulatory objective	Adherence to standards (1:100 year floodplain; 1:20 year floodway).	Achieving goals - balancing the social, economic and ecological costs and benefits of floodplain regulation.
Regulatory approach	Deterrent emphasis	Cooperative emphasis
Intergovernmental linkages	State standards; joint monitoring; local enforcement, with provision for state action.	State guidelines; local refinement of standards; local enforcement .
Implicit assumptions	Strong state role required to achieve regulatory goals; commitment and/or capacity weak.	Local governments are committed to regulatory goals; some require assistance to increase capacity.
Instruments	*Capacity-related* * Flood mapping (some joint activity) * Technical assistance for flood delineation * Limited funds available. *Commitment-related* * Prohibition on state-assisted or funded development in flood-prone areas where flood-free alternatives are available. * Potential legal liability for allowing flood prone development.	*Capacity-related* * 'Floodplain development manual' sets out guidelines * Technical assistance for local flood and flood management studies * Some funding for project implementation * Creation of local floodplain management committees *Commitment-related* * Tax incentive for leaving flood-prone land vacant. * Liability waivers for flood-related actions or decisions consistent with state policy.

Source: May and Handmer (1994) and May et al (1996, p. 76).

Coercion

In a coercive approach, higher level governments set out detailed standards and procedures to be implemented by local governments to achieve policy goals (Table 21.1). In effect, the lower level government becomes an agent following specific instructions from above, which allow little local discretion or room for innovation. The emphasis is on the need to satisfy detailed prescriptions rather than on broader performance criteria (Table 21.2). Coercion comes from mechanisms for monitoring the actions of local government, and in the form of penalties for failure to comply. In Florida for example, local jurisdictions can be, and are, fined for failure to comply with a coercive hazard management regime (May et al, 1996). The approach presumes conflicts between the various levels of government over goals or over the means for meeting those goals. It concentrates on building, or rather forcing, commitment instead of capacity. Commitment under these circumstances is likely to be of the "calculated" variety, intended to just meet the legal requirements and avoid the sanctions. This may become evident for example in the quality of floodplain management plans.

Limitations with coercion stem from the need for an institutional structure capable of monitoring compliance with the policy, and with taking effective enforcement action when compliance is lacking. Among other things this requires the availability of penalties and the means and will to enforce them. Even when the necessary bureaucratic and legal institutions and power exist - and their existence is by no means universal - they may be very difficult to apply. Heavy handed enforcement is often greatly resented and resisted, and may create a political backlash which threatens the whole policy (see Handmer, 1985). In reality, strong local opposition will often lead to negotiated solutions even in a coercive policy environment.

Attributes

Whatever approach to policy design is adopted, the local attributes of capacity and commitment are required.

Capacity

Although local implementation capacity is required for both cooperative and coercive approaches, the emphasis varies. As mentioned earlier, cooperative policy designs emphasise capacity building, while more coercive approaches tend to assume that the necessary capacity exists and concentrate on ensuring commitment. Organisational capacity is concerned with both what an organisation is doing, and the results it is achieving. However, most work on the topic stresses the means, or process, aspect; this is because a "capable" organisation has the ability or capacity to achieve a wide range of results (Honadle, 1981).

In practical terms capacity may refer to, among other things, ability to obtain funding, technical assistance, and training and development. It also refers to the ability to learn, to negotiate, to mobilise support for its objectives, and to possession of an adequate legal framework. The funding aspect is easily overplayed - in that organisations often lack the ability to absorb substantial increases in resources. It is not about the development of a perfect administrative system as this would ignore politics and the increasing societal pressure to negotiate outcomes among stakeholders.

Capacity building can be considered at a number of levels from the in-house capacity to undertake technical studies and access to legal advice, the place and power of planning within local government structure, to the conventions governing information flow across government bodies. One issue is whether capacity for hazard management is weakened or strengthened by the commercialisation of government activities (Hood and Jackson, 1992). Local authority trends in Europe are likely to have significant impacts on capacity. More capacity to undertake flood hazard management is likely as local entities expand, or form networks of common interest. But more capacity and different types of capacity will be needed if satisfactory outcomes are to occur in an increasingly deregulated competitive context - the competitive entrepreneurial skills needed to attract resources and development are being emphasised, but specialist expertise is needed for planning to minimise flood hazards.

Potential indicators of capacity emerging from current research can be characterised loosely as: power or authority, availability of money and expertise, and size in terms of population - with population often being a satisfactory surrogate measure for funds and expertise. A Council of Europe report (n.d.) suggests that local governments with fewer than about 5,000 people lack the capacity to undertake many tasks effectively. This affects many European countries, although compensatory mechanisms are increasingly being employed such as cooperation between communes and the use of private consultants (Marcou, n.d.). In addition, some countries lack capacity because there is no tradition of local planning.

In summary, capable organisations are forward thinking, learning, adaptive, networked with other organisations, politically astute, and able to solve problems.

Commitment

Local governments may have the capacity to implement flood hazard management programmes, but see them as low priority for a variety of reasons. They may believe that there is not a local flood problem; they may be unwilling to cooperate because of perceived problems with the policy or with other organisations or individuals involved, for example they may feel that they lack the necessary legal authority; they may be fully absorbed dealing with other local problems; they may be under pressure to allow development to proceed unheeded in flood prone areas, or they may have no support from their constituents. Often, this attitude is understandable, as floods seem to be rather remote low probability events compared with other immediate demands for attention or pressure for development. Although here I have separated commitment from capacity, in practice they are often closely related.

Lack of commitment is a serious problem that can undermine otherwise excellent policies or lead to total failure. The cooperative type policies tend to assume that commitment to the policy objectives exists. For the reasons set out in the paragraph above, commitment may often not exist, with the result that the performance of some local governments, in terms of flood hazard management, may be poor and decline steadily. In contrast, the coercive approach assumes that commitment is lacking, and works to create it. An obvious way of encouraging commitment is to make flood hazard management a legal requirement, with penalties for non-compliance. Commitment may also emerge through the professional standards or expectations of local government engineers and planners. Again, size may be important here. A report by the Council of Europe indicates that smaller jurisdictions tend to lack professional staff (n.d.: 35). Those local jurisdictions without strong commitment to flood hazard management may have a relatively minor flood problem. Unfortunately, a major problem may evolve in the absence of appropriate hazard management, if for example, substantial unconstrained development occurs in flood prone areas. In NSW, policy compliance was low among those authorities who regarded their flood risk as low. Information targeted at government officials or at the "public" may help build commitment. Such information may be drawn into a formal planning process designed to develop a strategy for flood hazard management. A requirement that local government undertake planning can help build local commitment and capacity (Burby and Dalton, 1992; Burby et al, 1996).

The evidence is that in Europe commitment to economic development is taking priority over environmental, natural hazards, and social issues at the local level. This jostling for economic development is at least partly a consequence of the enhanced economic competition resulting from European integration (Healey, 1994; Healey et al, 1995; Council of Europe, n.d.).

Conclusions

Central government strategies are also being put in place to manage flood hazards in many jurisdictions. However they must compete with other national and local priorities - many of which may seem far more worthy. Local capacity to deal with many problems may be increasing, but the increased capacity seen may not be appropriate for hazard management. To be effective, both cooperative and coercive approaches require degrees of commitment and capacity. In particular, coercive strategies need an institutional structure capable of monitoring compliance with the policy, and with taking effective enforcement action against non-compliers. Even when the necessary bureaucratic and legal institutions and power exist - and their existence is by no means universal - heavy handed enforcement is often strongly resented and resisted.

A cooperative approach to policy design and implementation offers an avenue which greatly reduces the need for monitoring and enforcement by higher levels of government, and acknowledges that to be fully appropriate and long lasting solutions need to be tailored to local circumstances. Instead of an enforcement mentality, flexibility, adaptability, and a willingness and ability to negotiate are required - characteristics which cannot simply be legislated. The psychological need is to keep the ultimate goal of flood hazard management in sight; rather than simply being locked into enforcement of detailed regulations. The resource needs are for capacity building. An important aspect is the establishment of participative processes for discussion and negotiation between the various stakeholders.

Our research demonstrates that when local governments lack commitment to higher level policy objectives, a coercive approach produces better results as measured by local effort and compliance with specified procedures (May et al, 1996). But where commitment already exists, a cooperative approach leads to results equal or better than those achieved under coercion. In addition, it appears that cooperative policies may be superior in maintaining local government commitment. The remaining problem then is obtaining commitment from those local authorities who view flooding as unimportant. Possibilities include building capacity, and prescribing and supporting a planning process - it was found that a process which includes major stakeholders tended to help build greater commitment.

Acknowledgements

This paper draws on ideas developed in a collaborative research project, with substantial support from US National Science Foundation grant BCS-9208082 administered by the University of Washington, Seattle. The support of the directors and staff at the Centre for Resource and Environmental Studies, Australian National University, and at the Flood Hazard Research Centre, Middlesex University, is also appreciated. The project involved interviews with government officials, questionnaire surveys of local authorities and central government in three jurisdictions: Florida (USA), New South Wales (Australia) and New Zealand. In addition, case studies were undertaken in NSW and New Zealand. The results of the full study are published in May et al (1996).

Note

1. This is a shortened and modified version of a paper by John Handmer published in 1996 as 'Policy design and local attributes for flood hazard management', in the *Journal of Contingencies and Crisis Management*, vol. 4, no. 4, pp. 189-197, published by Blackwells.

References

Bollens, S.A. (1993), 'Restructuring land use governance', *Journal of Planning Literature*, vol. 73, pp. 211-226.

Brouwer, O.W. (1994), 'Subsidiarity as a general legal principle', in Dubrulle, M. (ed), *Future European Environmental Policy and Subsidiarity*, European Interuniversity Press, Brussels, pp. 27-43.

Burby, R.J. and Dalton, L.C. (1992), 'Plans can matter! The role of land use plans and state planning mandates in limiting development of hazardous areas', *Public Administration Review*, vol. 543, pp. 229-237.

Burby, R.J., May, P.J., with Berke, P.R., Dalton, L.C., French, S.P. and Kaiser, E.J. (1996), *Making Governments Plan: State Experiments in Managing Land Use*, Johns Hopkins University Press, Baltimore.

Council of Europe (not dated), *The size of municipalities, efficiency and citizen participation. Local and Regional Authorities in Europe*, no. 56, Brussels.

Handmer, J.W. (1985), 'Flood policy reversal in NSW, Australia', *Disasters*, vol. 94, pp. 279-285.

Handmer, J.W. (1996), 'Policy design and local attributes for flood hazard management', *Journal of Contingencies and Crisis Management*, vol. 4, no. 4, pp. 189-197.

Healey, P. (1994), 'The issues', in Healey, P. (ed), *Trends in development plan-making in European planning systems*, Department of Town and Country Planning, University of Newcastle, Working Paper no. 42, pp. 5-18.

Healey, P., Cameron, S., Davoudi, S., Graham, S. and Madani-Pour, A. (1995), *Managing Cities: The New Urban Context*, Wiley & Sons, Chichester.

Honadle, Beth W. (1981), 'A capacity-building framework: a search for concept and purpose', *Public Administration Review*, September/October, pp. 575-580.

Hood, C. and Jackson, M. (1992), 'The new public management: a recipe for disaster?' in Parker D. J. and Handmer, J.W. (eds), *Hazard management and emergency planning*, James and James, London, pp. 109-125.

Marcou, G. (not dated), 'Intermunicipal cooperation as a means to improve the efficiency of local authorities', in Council of Europe, *The size of municipalities, efficiency and citizen participation. Local and Regional Authorities in Europe no. 56*, Brussels, pp. 141-152.

May, P. and Handmer, J.W. (1992), 'Regulatory policy design: cooperative versus deterrent mandates', *Australian Journal of Public Administration*, vol. 511, pp. 43-53.

May, P. and Williams, W. (1986), *Disaster Policy Implementation: Managing Programs under Shared Governance*, Plenum Press, New York.

May, P., Burby, R.J., Ericksen, N.J., Handmer, J.W., Dixon, J.E., Michaels, S. and Smith, D.I. (1996), *Environmental Management and Governance: Intergovernmental Approaches to Hazards and Sustainability*, Routledge, London.

22 Decentralisation, Local Autonomy and Metropolitan Influence: The Case of Salvador, Brazil

CELINA SOUZA

Introduction

Salvador is the main city of the Salvador Metropolitan Region (SMR). The SMR comprises ten municipalities and covers an area of 2,181 square kilometres.[1] The region's population in 1995 was 2,496,521, being the fourth largest city in Brazil. Salvador has the role of being the urban and services centre for business set up outside its territory and a focal point for skilled workers. Because industrial activities have not flourished in the capital, this makes the SMR an exception among Brazilian metropolitan regions.

The political features of the SMR were assessed for the first time in the early 1990s in a survey involving regional leaders (Medeiros and Souza, 1993). The survey showed that the SMR was marked by contradictions between a 'new' rationality brought about by industrialisation and urbanisation as against features of the 'old' agrarian rationality. The coexistence of these contradictory rationalities indicates that a new balance of power, redefining social and political relations, is in the making as a result of economic modernisation.

This paper investigates the effects of decentralisation and of the role of non-elected metropolitan authorities upon local political autonomy and environmental management. The municipality of Salvador is taken as a case study to illustrate the discussion as whether political and financial decentralisation brought about by democratisation has improved local political autonomy and local environmental management. The article argues that decentralisation improves democracy through the incorporation of new political groups to the local scene. However, in the specific case of Salvador, decentralisation has blocked the possibilities of financial help from the federal government, thus aggrandising the financial and political role of metropolitan authorities in municipalities which shelter the very poor and which have no important economic activity. In this case, the implementation of environmental policies has to be supported by the metropolitan authority, therefore limiting the scope for local elected politicians to play a more active role in any kind of environmental management.

The paper is structured as follows. Section 1 presents a conceptual discussion of political autonomy and its relation with environmental management. Section 2 analyses the municipality of Salvador, showing its political importance in contrast to its financial weakness. Section 3 provides some key information about the Salvador Metropolitan Authority, *Companhia de Desenvolvimento da Região Metropolitana de Salvador* (CONDER), and analyses the waste processing and disposal programme implemented by this institution.

Local Political Autonomy and Environmental Management

Local political autonomy can be seen in various ways. Financially, Brazilian municipalities were the main beneficiaries of decentralisation introduced by the political opening of the 1970s and by the 1988 Constitution. Furthermore, they exercise a significant discretion in allocating their own revenue, and for administering their resources. They are also highly capitalised as compared to their counterparts in developing countries. In 1992 it was estimated that no municipality had received less than US$400,000 a year from the Municipal Participation Fund, the FPM (Bremaeker, 1994, p.18), whereas in Latin America a great number of municipalities have an annual budget of between US$5,000 and US$20,000 (Lordello de Mello, 1991, p. 215). Revenue transfers by the federal government alone amounted to US$5.5 billion in 1993 (Bremaeker, 1994, p. 37).

Despite this financial strength, not all Brazilian municipalities are in a more favourable position following decentralisation. Bremaeker (1994) showed that more than 200 municipalities in the Northeast have no chance to levy their own revenues. The reason for this is the lack of industries and/or the size of their poor population, which shows that they cannot survive without the financial help of other levels of government and/or other institutions such as the metropolitan agencies. Brazil's experience displays the limits of decentralisation in countries characterised by regional inequalities even when they enjoy a fiscal mechanism designed to address the issue of intra- and inter-regional disparities as in Brazil.

The second way through which political autonomy can be seen is the political-juridical viewpoint. Since the 1988 Constitution the municipalities are members of the federal pact, together with the states, a unique feature among federal countries. Furthermore, their revenues and jurisdictions are constitutionally assured, meaning that changes in these issues can only be made through amendments to the constitution.

Political autonomy can then be defined as the capacity of local governments to settle and accomplish, to a significant degree, their own political agenda and their own public policies concerning issues that belong to the local jurisdiction. Local political autonomy is used here to refer to the existence of political and administrative room that might be used by local political forces once they are elected to executive positions. The autonomy to settle a political agenda and to implement public policies would be relatively free from the influence of political parties or pressure groups that do not belong to the political coalition elected to administer the city.

The assumption made here is that with democratisation and the increase in local revenues and jurisdictions it is possible to change the political agenda and the management at the local sphere in the Brazilian municipalities, although to differing degrees. Therefore, local autonomy is likely to improve the role of local governments in environmental management programmes. Environmental management in this context includes, among other things, the initiative to pass local legislation, the existence of local institutions managing environmental programmes and the participation of local actors (elected politicians, practitioners, NGOs and financial bodies) in environmental policies and management.

The Case of Salvador

As already noted, Salvador is the third most populous city in Brazil, after São Paulo and Rio de Janeiro, with some 2.3 million inhabitants. The city has accumulated a number of unresolved problems. In 1991 only 23% of the population had sewerage services; less than half had access to a refuse collection service; 39% were illiterate; 300,000 were unemployed and 80,000 were street peddlers. Despite these figures, Salvador is an important tourist destination: each summer around 400,000 tourists visit the city (*Folha de São Paulo*, 1992).

Financially Salvador is Brazil's most deprived state capital as a result of the lack of industry and the size of its poor population. The revenue structure reflects the city's dependence upon revenue transfers: less than 40% of the city's revenue comes from its own taxation.

Since 1988 most state capitals have increased their revenue not only in the amounts transferred from the federal and state governments, but through an effort to increase their own revenues and to adjust their finances, but not Salvador. The city's chances of overcoming its financial problems were blocked by several pressures. One of them was a judicial sanction imposed on the city's monthly revenue to pay its debt with the construction industry. Estimated at US$200 million, this debt was to be due to be paid by the year 2014. In 1991 seven construction companies won the approval in court of an agreement which paid them 20% of the monthly quota of the city's FPM. In 1996 the courts eventually decided to lift the sanction, recognising that the city authorities had paid more than its outstanding debt to the construction industries.

Another problem is the relationship between the city's debt and its net revenue. Salvador's debt rose from 37% in 1986, through 200% in 1987 and 220% in 1988 to 453% of the city's net revenue in 1989. Currently (1998) the city's debt amounts US$421 million, of which US$402 million is in short-term debt, implying very high interest rates.

These problems result in Salvador's revenue collection accounting for just 2.5% of total revenue collected in state capitals, being considerably less than cities with a smaller populations (ABRASF, 1992). As revenue collection does not improve, the city has been using two perverse ways to pay its payroll and other regular bills. One is to borrow from private commercial banks at market interest rates, which are high in Brazil. The second is to ask for the 'good will' of the state governor in anticipating the city's quota from the ICMS (Value-Added, Communication and Transportation Tax). The former is partially responsible for the increase on debt. The latter means that mayors have to rely on the governors for cash, which implies that they become dependent upon the governors to accomplish the city's ordinary duties.

Political Resources[2]

The first mayor elected in Salvador by popular poll after the political opening in 1986 was Mário Kertész, then affiliated to a centrist party, the PMDB (Party of the Brazilian Democratic Movement). In 1989 Fernando José, a former television football broadcaster, was elected. He was also affiliated to the PMDB but changed party during his term. In 1992 the electorate chose Lídice da Mata, from the PSDB (Party of Brazilian Social Democracy).

Despite belonging to different political groups, Salvador's mayors have shared a common feature as a result of the city's financial weakness: they have little room to take their own decisions and to define their own agenda. The city's politicians are forced to ask for the support of other groups who were not elected to govern the city, such as the state government, the construction industry and the metropolitan authority. His burden has sealed the destiny of Salvador's politicians: to lose control of the city's decisions and to become prisoners of their sponsors' will.

As for environmental management, Salvador does not have a specific institutional structure to deal with environmental issues. The local Planning Secretariat is responsible for pollution and noise control because the industries that are the main source of environmental problem are located outside Salvador. The state of Bahia has an agency, the CRA (Centre of Environmental Resources) linked to the state Planning Secretariat which is responsible for minor environmental programmes and for the licensing of activities that may harm the environment.

Local and state agencies have no jurisdiction in programmes such as refuse collection and disposal, being involved only in routine operations. Refuse collection and disposal was, therefore, in a governmental limbo and there was no struggle among local agencies for its control. Also non-governmental environmental groups concentrate their efforts in pollution control and in saving what is left of the state's original forests along the Atlantic coast.

To sum up, contrary to the predicament of other state capitals, Salvador did not benefit from political and financial decentralisation. The lack of resources has meant a stronger dependence upon the state government and the construction industry. Salvador is an exception to the rule that financial decentralisation has had positive effects in most state capitals. Furthermore, financial decentralisation has proved to have little effect on improving the mayor's capacity to govern, given Brazil's high inter and intra-regional disparities.

Salvador's Metropolitan Authority and Its Environmental Programme

CONDER is a state enterprise linked institutionally to the Planning Secretariat of the Government of Bahia, and is responsible for the co-ordination and implementation of the development policy of the SMR. The company was set up in 1974, following the creation, by the military regime, of Brazilian metropolitan regions under the state jurisdiction but with federal financial backing. CONDER has always enjoyed great prestige and has been able to levy substantial resources, including those from international financial agencies, making it an exception among metropolitan authorities, which, in general, have always had low prestige and few resources.

Several reasons have contributed to CONDER's success. Firstly, during the military regime all but one of the mayors of the SMR were appointed by the state legislature. The members of the legislature are traditionally dependent upon the state governors' support such that the mayors were inevitably men known to support the policies and to go along with decisions of the metropolitan region that was run by non-elected representatives.

Secondly, the military regime centralised financial resources making transfers mainly to the states and to some metropolitan authorities according to the regime's economic and political goals. Therefore, the existence of a technocratic-oriented company, CONDER, as opposed to low-skilled municipal staff, inclined the regime to make considerable federal financial resources available to the SMR via CONDER.

Thirdly, Bahia's political life has, since the military regime and after the political opening, been controlled by a charismatic and authoritarian leader, the above mentioned Magalhães. Magalhães had always shown respect for technocratically oriented quasi-government institutions. Fourthly, because Bahia's political leaders have, since economic modernisation, insulated the planning and economic practitioners from patronage and clientelism, CONDER's officials have traditionally been well paid and adopted a sense of mission and a commitment to CONDER's objectives.

However, the most important event leading to the upgrading of CONDER's participation in the SMR was the establishment of a direct relationship with the World Bank through a programme called simply the Metropolitan Project. Between 1983 and 1997 the company had a contract with the World Bank amounting to US$77 million aimed at the development of various urban infrastructure and maintenance programmes that have included a waste water management project, a solid waste collection and disposal project and a sanitary education programme.

The solid waste management project is the most important and expensive project within the Metropolitan Project, accounting for 41% of the investment. On their own, given their lack of revenue-raising capacity, none of the SMR municipalities - not even Salvador - would have qualified to obtain finance from the World Bank. But it also true to say that neither local environmental agencies nor environmental pressure groups have any experience in dealing with urban environmental programmes such as solid waste collection and disposal.

Tackling all aspects of waste and its disposal poses a challenge to most Brazilian cities. Data show that only 3% of Brazil's urban refuse is adequately disposed of (Mercado, 1996). Open refuse disposal accounts for 88.2% of all refuse collected in Brazil's almost 5,000 municipalities. The SMR is no exception. In 1992 84.8% of the waste was disposed of in several large open dumps. These open dumps received 2,227 tonnes/day of waste. The municipality of Salvador collected only 70% of the garbage it produced. In addition, the productivity of the system is extremely low.

In street-sweeping, for instance, the recommended level of productivity is 1,500 meters per worker per six hour shift, but the region's average was 765. As for the relationship between the number of workers and the waste they collect, the recommended level is one worker/tonne per day but the regional average in 1992 was 5.2 worker/tonnes, showing a huge waste of financial resources (CONDER, 1996).

While CONDER sponsored three environmental programmes with the financial backing of the World Bank, the municipality of Salvador had none, except its regular duty of waste collection and street-sweeping. Within CONDER's programme the major item of expenditure is the construction of four sanitary landfill sites which will replace the region's open dumps. Three landfill sites are in operation while one is under construction. The landfill sites are administered by CONDER with no participation of local authorities, local financial institutions or local NGOs.

Therefore, there was no room for improvement of local environmental management programmes, given that no local legislation was passed, there are no local institutions in charge of managing environmental programmes and no local actors are involved in managing or financing environmental programmes. For environmental policies to be implemented they have to be supported by institutions which do not have a mandate to define the local agenda, primarily in the form of Salvador's metropolitan authority. This situation is a major disincentive to local elected politicians, as well as other local actors, to playing a more active role in any kind of environmental management. This picture also shows that there are a number of unresolved problems which have to be tackled outside the local level which prevent local governments from playing an active role in environmental programmes.

Conclusions

One paradox emerges from the case of Salvador: the political power and financial decentralisation which accrued to the city through the 1988 Constitution were insufficient to free Salvador's politicians from dependence on state political leaders, its metropolitan agency and the construction industry. The effects of this in Salvador must be stressed.

Firstly, economically, Salvador benefited from economic deconcentration from the Southeast because of its role as an urban centre for business and skilled workers. Nevertheless, these changes were not enough to cope with the size of the population which has historically been cut off from the benefits of economic development. The gap between economic results and the conditions of the majority not only penalises the poor but also limits local government action. Local politicians have to give up their role of improving the living and environmental conditions of the population in the long-term, leaving these issues to the will of their sponsors.

Secondly, in terms of policy priorities, as the mayors are obliged to attempt to mark their administration with some achievement, they opt for projects with short-term political pay-offs. Therefore, projects requiring long-term maturity and/or having low visibility to the public - such as environmental projects - are ignored.

Third, in intergovernmental relations, the state government has always had a strong presence both in Salvador and in its metropolitan region, being, for the most part, the major decision-maker. However, as a branch of the literature on local government suggests, local politicians are closer to communities and hence able better to identify and respond to their needs. As the municipality analysed is made up of basically of poor constituents and has few financial resources, the mayors are hindered from providing their constituents with access to public services or to an improved environment. Given the high costs involved, the metropolitan authority is the only agency with the means to embark on environmental management projects.

Although in theory decentralisation fosters democracy by broadening the number of participants in the local political arena, in the case of Salvador there are political and economic factors at work which largely predetermine outcomes and hence vitiate any real increase in democracy. This situation would clearly apply to any local government which possesses either a large poor population and/or is with a weak economic base.

Financial decentralisation and democratisation were commended by the politicians to the people as ways to improve, among other things, their access to better standards of living in which environmental issues are important. However, this paper has shown that local politicians have little room to implement environmental programmes in cases where their constituencies shelter the very poor. This finding stresses the limits of decentralisation in countries characterised by deep-rooted intra- and inter-regional and social inequalities, showing that there are unresolved problems which need to be tackled beyond the local level before mayors can play an active role in environmental management and in improving their constituents' living and environmental conditions.

The case of Salvador also shows how policies such as decentralisation cannot be seen as the panacea for problems which are larger than their possible solutions. Decentralisation in Brazil has contributed to the development of democracy by the incorporation of several different political groups within the local scene but its impact upon public policies in general and environmental policies in particular is limited, even in cities of regional and national importance such as Salvador.

Data have shown that decentralisation and democratisation have brought about a fragmentation of power without changing the implementation of environmental policies locally. Following this assertion, one can conclude that there are political and economic factors influencing the results of decentralisation and that in certain urban areas environmental policies and management have to rely more on metropolitan institutions financed by international organisations rather than on local structures.

Notes

1   All Brazilian municipalities have the same legal status. The official definition of a municipality, or *municipio*, encompasses two levels of local government which may be urban or rural.
2   A more detailed analysis is presented in Souza (1992; 1997).

References

ABRASF, Associação Brasileira dos Secretários e Dirigentes das Finanças Municipais das Capitais (1992), *Relatório*, (several reports), unpublished.

Bremaeker, F. (1994), *Mitos e Verdades sobre as Finanças dos Municípios Brasileiros*, IBAM, unpublished.

CONDER, Companhia de Desenvolvimento da Região Metropolitana de Salvador (1996), *Indicadores de Limpeza Urbana: Região Metropolitana de Salvador*, CONDER, Salvador.

Folha de São Paulo (1992), 7 April.

Lordello de Mello, D. (1991), 'Descentralização, Papel dos Governos Locais no Processo de Desenvolvimento Nacional e Recursos Financeiros Necessários para que os Governos Locais Possam Cumprir seu Papel', *Revista de Administração Pública*, vol. 25, no. 4, pp. 199-217.

Medeiros, A C. de and C. Souza (1993), 'Gestão do Território *Versus* Estrutura de Solidariedade e Autoridade', *Revista de Administração Pública,* vol. 27, no. 3, pp. 37-49.

Mercado (1996), December.

Souza, C. (1992), 'Political and Financial Decentralisation in Democratic Brazil', *Local Government Studies*, vol. 20, no. 4, pp. 588-609.

Souza, C. (1997), *Constitutional Engineering in Brazil: The Politics of Federalism and Decentralization*, Macmillan, London and St. Martin's Press, New York.

23 Controversy over the Preservation of a Metropolitan Area's Fresh Water Reservoirs: Legal Instruments and the Politics of Environmental Management in Istanbul, Turkey

AYSE YONDER

Introduction

Over the past decade there have been numerous stories in the Turkish media about 'illegal cities' that have developed through the activities of the 'land Mafia' on public and forest land near Istanbul's fresh water reservoirs.[1] In fact, population living within the metropolitan water reservoir areas increased from 190,000 in 1985 to 466,600 in 1990, growing at a much faster rate than the city average[2] (Uysal, 1995, p. 52). Only in 1996, the Istanbul branch of the Chamber of Architects of Turkey filed two lawsuits against the Metropolitan Municipality of Istanbul. The first, filed on February 16, 1996 against the Metropolitan Municipality and Istanbul Waterworks and Sewage Authority (ISKI), was to annul and stop any action based on the new ISKI regulation of December 26, 1995, which introduced significantly lower water pollution control standards than the previous pieces of legislation.[3] The second lawsuit was filed to annul and stop implementation of specific items of the master Plan for Istanbul Metropolitan Area, adopted by the municipality on October 20, 1995. The purpose of both lawsuits was to prevent legitimisation of existing and future development around the city's fresh water reservoirs.

261

Both actions also reflect the concern that the new environmental legislation would be used in the same way by the local governments as the land development legislation. The rapid growth of Istanbul since the 1950s has been accompanied by a speculative boom in urban property markets, which benefited almost all social and economic groups until the late 1970s. The clientelist nature of party politics at the local and national levels fuelled both the speculative building boom and informal settlement formation. While large scale, powerful real estate interests could mediate variations in zoning and development regulations through top-level officials directly and individually. Negotiation and relaxation of controls was critical, however, in the formation of low-income settlements. At the local level, municipal governments, responsible for delivery of infrastructure services, and enforcement of land use and building regulations, lacked the financial or technical capacity to undertake these functions (Yonder, 1982). Selective enforcement and relaxation of regulation thus became 'the most expedient way of dispensing patronage' in return for votes in both the formal and informal districts of the city (Oncu, 1987, p. 45).

The purpose of this paper is to discuss issues related to environmental management and the politics of development around Istanbul's fresh water reservoirs within the political and institutional context of the 1990s. After a brief discussion of Istanbul's changing growth pattern and the developments around the city's water reservoirs, I will review the main features of the current environmental legislation and the metropolitan government system in Istanbul. I will consider the problems and potential of the political and institutional context in Istanbul in terms of sustainable development of the metropolitan area.

Istanbul's Changing Growth Pattern and Development Around the City's Fresh Water Reservoirs

Istanbul is the largest metropolitan centre in Turkey and accounts for 20 percent of the total urban population. As the leading harbour and major industrial, financial and cultural centre, Istanbul's population increased rapidly since the 1950s, from less than a million people in 1950 to three million in 1970, to over seven million in 1990. It is estimated that about ten to twelve million people now live in the Istanbul metropolitan area.

Historically, settled areas in Istanbul were confined to a narrow strip along the coast of Marmara Sea, and the Bosphorus. Forest areas protecting the city's precious water reservoirs, and lack of roads restricted growth further inland. Until the late 1970s, Istanbul grew through density increases and middle to high-income development along the coast, and low-income settlements further inland, near the industry located along the E-5 highway. With the construction of the first Bosphorus Bridge in the early 1970s, a similar pattern of growth started on the Asian side of the city.

During the 1980s, structural adjustment measures and new public infrastructure investments led to transformations in the city's growth and socio-spatial segregation patterns.[4] First, manufacturing industry (90 percent private investment) which until the late 1970s had been concentrated in Istanbul, started to disperse and locate elsewhere in the region in the 1980s. Istanbul's share in the country's manufacturing value added declined from 34 percent in 1975 to 27 percent in 1991.[5] Yet, there is a significant concentration of small industry in Istanbul, and substantial numbers are located within the city water reservoir areas. According to Eroglu, Sarikaya and Sevimli, only 532 of the 1,420 industrial establishments (approximately 37 percent) that are members of the Istanbul Chamber of Industry have waste water treatment facilities (1996, p. 8).

Second, an increasing number of upscale housing developments started to locate inland, within the outer ring, formerly populated only by low-income groups. The northward expansion of both new high and low-income settlements started to threaten the city's limited fresh water resources. The new upscale developments, popularly called 'site', are often in the form of gated communities, and represent new forms of segregation even among the middle and upper-income groups themselves. The new express ring road and highway system directly connects the site's to new high-rise office buildings, shopping malls, and the city centre. Hence, even though different income groups now live in greater spatial proximity to each other than before, their contact is minimised (Yonder, 1998). Meanwhile, densities increased in old informal settlements and new high rise-high density informal settlements started forming near the city water reservoirs, such as Sultanbeyli near the Omerli reservoir. The prevalence of apartment buildings in informal settlements was one of the factors that started to change the perception of the general public towards the residents of these areas in the 1980s, viewing these processes as oriented to speculation rather than being based on need as before (Sen, 1996, p. 21). The swing in the votes of low and moderate income groups, mostly from informal settlements, from left-of-centre in the 1970s to extreme-right parties in the 1990s added a political dimension to the social and economic polarisation.

As Table 23.1 shows, both industrial and residential developments create sources of pollution within the metropolitan water reservoir areas. According to Uysal, a total of 19 new district municipalities have been established within the six water reservoir areas from 1985 to 1995 (Uysal, 1995, p. 54). Population increase has been the fastest especially around Omerli and Elmali reservoirs on the Anatolian side of the city and around Alibeykoy reservoir near the old industrial zones on the European side (Uysal, 1995, p. 52). According to Budak and Tuzun, the number of industrial enterprises located in Istanbul's seven fresh water reservoir areas increased from 848 in 1989 to 1,352 in 1993. In 1990, ISKI identified 253 enterprises that had to be moved out of these Special Protection Areas (1993, pp. 47-48).

Table 23.1 Residential and industrial developments around Istanbul's fresh water reservoirs, early 1990s

Reservoir location, name and year of construction	Area (km^2)	Capacity (Million m^3)	Share of city's water supply (%)	Population and industrial activity in catchment area	
				Population	No. of manufacturing firms
European side					
Terkos Lake					
1883- 1950)	619	238	22.2	33,250	22
Alibeykoy Dam (1972)	160	66	6.6	62,000	380
B. Cekmece Dam (1989)	620	162	17.0	103,500	208
Anatolian side					
Omerli Dam (1972)	621	387	31.8	217,500	454
Darlik Dam (1989)	297	110	16.5	6,950	na[a]
Elmali Dam (1950)	81	12	2.6	122,500	143
Other freshwater sources[b]	--	na	3.4	1.1	na

a. Not available
b. Wells, historic cisterns, small reservoirs
Sources: Budak and Tuzun (1993), and Istanbul Metropolitan Municipality (1995).

The new developments have been facilitated by major infrastructure investments and planning decisions by central government agencies. The new ring expressway cuts through both Elmali and Omerli reservoir areas, and there are new planned industrial zones within both reservoir areas. These include the Dudullu industrial zone in Omerli and the industrial zone allocated for the relocation of leather industry to the south of Elmali reservoir. The implementation of a central government decision to locate an Industrial Free Zone in Catalca and another decision to open 1,000 hectares to residential development near Elmali reservoir were stopped through law-suits of the Chamber of Architects of Istanbul. Local master plans that do not match the actual development within some of the district municipalities, and the fact that the new district municipalities near the reservoirs have been left out of the Metropolitan Master Plan decisions are other areas of concern emphasised by the Chamber of Architects.

Metropolitan Government System and Institutional Framework for Fresh Water Resources Preservation in Istanbul

Davey (1993) identifies four conditions that are necessary for local governments to provide effective leadership in the overall development of their jurisdiction and in responding to respond to economic, social, and environmental problems. These are:

1. comprehensive boundaries including the developed and developing periphery;
2. wide responsibilities around the functions that interrelate constantly during rapid urban growth, i.e., planning and development control, water supply and sewage, roads and traffic management, drainage, regulation of public transport, parks and open spaces, and environmental health;
3. a buoyant revenue base, including access to taxation of income or expenditure; and
4. well-qualified staffing (Davey 1993, p. 47).

A series of urban reforms in the 1980s increased the planning, enforcement and services provision capacity of local governments and improved the condition of local governments in the above mentioned areas. First, the metropolitan government system was reorganised in 1984. A two-tier municipal government system was established with a metropolitan municipality and 15 district municipalities.[6] The jurisdiction of municipalities was expanded to control new development areas. The financial capacity of municipalities was significantly improved and a cost recovery system was adopted in service provision in 1981. All water and sewage services were centralised under the semi-autonomous Istanbul Water and Sewage Authority (ISKI) of the Metropolitan Municipality, with a substantial budget through local revenues and international credit.

Improvements in the local revenue sources of municipalities aimed to reduce their dependency on the central government (Laws no. 2380 and no. 2464 of 1981). A major step in improving municipal revenues was to charge them with the collection of property taxes within their jurisdiction. Moreover, a real estate tax reform in 1982 increased tax revenues, as well as the costs of property ownership. Finally, a new reconstruction law (no. 3192) was adopted in 1984. The new law decentralised the master plan approval from the ministry to the metropolitan level.[7]

It should be indicated here that these administrative reforms, along with a set of new housing legislation and institutions, also transformed the structure of the housing sector (Yonder, 1998). Most of the new measures were geared to and did stimulate formal sector production, and established a favourable legal, administrative and financial framework for large construction companies. Low-income housing policy, however, has continued to be *ad hoc* consolidation of existing informal settlements through a series of building amnesties since 1983 as before.

While there were a number of items related to environmental protection scattered in various pieces of legislation, it was in the 1980s that environmental issues started receiving more attention in Turkey. Article 52 of the new 1982 National Constitution declared a safe and healthy environment as a right of all citizens, and another (article 90) recognised international treaties ratified by the parliament as legislative acts valid as domestic law (Gonullu, 1996). It was after 1983 that comprehensive laws were introduced directly related to the environment.[8] The first comprehensive Environmental Law (no. 2872) was adopted in 1984. This includes the Ramsar Convention related to wetlands preservation adopted by the Turkish government in 1994. In 1991, the first Ministry of Environment was established to coordinate all environmental management and preservation responsibilities. The Turkish government signed the Rio Declaration and Local Agenda 21 in 1992.

Both the national legislation[9] and international agreements adopted as domestic law give the metropolitan and district municipalities a wide range of planning and environmental control responsibilities. Yet the legal tools for management and preservation of city fresh water resources, are still divided among a number of government agencies at different levels.

A number of issues have been raised at the seminars organised by the Chamber of Industry of Istanbul and ISKI around various aspects of environmental management and protection. These issues range from the insufficiency of bylaws and regulations to translate current laws into implementation, to inconsistencies in policies of different municipalities, to lack of coordination among the actions and requirements of different agencies and bureaucratic territorialism, which discourage permit applications among industrialists. Other issues relate to lack of sufficient staffing, funding and technical services at government agencies for monitoring activities and enabling prompt response to applications. Yet probably the most significant concern that is shared by industrialists, large real estate representatives, and professional and environmental organisations is the lack of 'respect' among state agencies of all levels to the laws of the state. This is demonstrated by the numerous revisions and amendments to environmental legislation over the past ten years. Another popular example of such practice is the development of the Suktanbeyli settlement near Omerli reservoir. In a recent issue of the ISKI Newsletter, one article claims the determination of ISKI to prevent all illegal construction within the city's water reservoirs. Two pages later, another article announces the opening ceremony of new fresh water facilities in Sultanbeyli, a new district municipality formed through informal processes.[10]

Yet, improvement of the existing environmental management practices cannot be viewed as a purely legal or technical arrangement. 'It is the local political processes make central policies politically sustainable, not technocratic, top-down solutions' (Cohen, 1995, p. 277). As a recent comparative study of institutional, policy and contextual factors related to farmland protection in six countries has shown, neither stringent legal controls nor any specific style of planning is sufficient to determine the level of success (Alterman, 1997). Rather, the successes of Netherlands and Great Britain can be attributed to other factors, such as widely shared 'norms for rights and obligations in land development' (1997, p. 231). In Israel, where 'norms of better countryside stewardship (were) ... too weak to meet (the) challenges and pressures of mass migration and development', CPAL, a well established central government institution was weakened at a time when most needed (Alterman, 1997, p. 235).

As the findings of a recent citizen survey show, environmental and infrastructure issues rank among the top citizen concerns in Istanbul.[11] Yet the environmental debates around the city's fresh water reservoirs do not include the voice of local citizens. Consensus around the definition of the environmental problems and priorities is critical especially in societies with a large informal sector. As Koksal indicates, when hundreds of thousands of citizens are involved in informal practices and transactions every day, we cannot expect top-down rules and regulations to be sustained socially.[12]

Conclusion

In Turkey there is now a much more favourable legal and institutional context for environmental management at the metropolitan level, compared to ten years ago. As discussed above, at least three of the four conditions indicated by Davey as necessary to achieve effective leadership at the local government level are accomplished. Moreover, public attention to environmental issues has also increased in the 1990s - at least among the middle classes, representatives of big industry and most civil society organisations.

Yet the risks to and the costs of preserving the city's fresh water reservoirs will continue to increase unless all levels of government, especially the municipalities, commit to an agenda of sustainable development and approach issues of land development, housing and environmental preservation in an integrated manner. Moreover, it is crucial to include all groups affected in the decision-making process. For municipalities and civil society organisations alike, this means organising to increase a sense of civic consciousness and relating "environmental issues" to people's own concerns and daily lives. Small industry and the residents of informal settlements, who are now all viewed as the polluters of the city's fresh water reservoirs but who themselves suffer daily from environmental problems, should be the main groups to involve in discussions to deal with this issue in a sustainable way.

Notes

1   For instance: E. Demircioglu, (August 16, 1988), 'Istanbul'da arsa mafyasi', *Milliyet,* p. 13.
2   These figures include Darlik, the seventh and newest reservoir area.
3   The standards introduced by the new 'ISKI Regulation on Protection of Current and Potential Surface Fresh Water Resources' contradicted those established by the national Water Pollution Control Regulations (April 9, 1988) based on the Environmental Law no. 2872 (August 11, 1983), and Law no. 2560 Concerning the Establishment and Functions of ISKI (November 23, 1984).
4   Interestingly, in a similar pattern described by Ribeiro and do Lago (1995) for Brazilian cities.
5   Still, headquarters of corporations are concentrated in Istanbul - about a half of the 500 largest firms in Turkey are members of the Istanbul Chamber of Commerce. The dominant sectors in Istanbul are now textiles, metal-automotive, and paper-newspaper, followed by chemicals, petrol and foodstuff (Sonmez, 1996, p. 46).
6   The number of district municipalities have almost doubled by now, with the establishment of new administrations on the periphery.
7   Some of the new items in the law include allowing the government to condemn up to 35 percent of the land for public uses during new plan preparation, and phasing out of the infrastructure requirement for large scale projects located outside of master plan boundaries.
8   These include the Environmental Law (no. 2872), Law for the Preservation of Cultural and Natural Resources (no. 2863) and National Parks Law (no. 2873).
9   Especially the Municipalities Law no. 1580 of 1930, Metropolitan Municipalities Law no. 3030 of 1984, and the Law no. 2560, Concerning the Establishment and Functions of Istanbul's Water and Sewage Authority.
10   *ISKI Haber (ISKI Newsletter) articles (*April 1996) 'Havzalari kirletenlere taviz yok', and 'Ilcenin altyapisi suratle tamamlaniyor: Sultanbeyli Icmesuyu Sebeke Insaatinin Temeli Atildi', no.2, pp. 16, 10 and 12.
11   International Republican Institute, 'Turkey, survey results: Attitudes and priorities of citizens of urban areas', 4/22-30/1995 and 11/9-23/1995, cited in T. Belge and O. Bilgin (1997), p. 109.
12   S. E. Koksal's comments in the panel debate organised by *Istanbul* journal on 'How should Istanbul be governed?', no. 8 (January 1994), pp. 152-3.

References

Abacioglu, M. (1995), *Cevre Kanunu ve Cevre Sagligi Mevzuati*, (Environmental law and public health legislation), Seckin, Istanbul.

Alterman, R. (1997), 'The challenge of farmland preservation: Lessons from a six-nation comparison', *Journal of the American Planning Association*, vol. 83, no. 2 (Spring), pp. 220-243.

Belge, T. and Bilgin, O. (1997) (eds), *Yurttas Kaltilimi: Silvi Toplum Kuruluslari ve Yerel Yonetimler Arasinda Ortaklik ve Isbirligi*, (Citizen participation: Partnership and collaboration between CSOs and local governments), Helsinki Citizens Assembly, Istanbul.

Budak, S. and Tuzun, G. (1993), 'Istanbul'da Icme Suyu Havzalari' (Fresh water reservoirs in Istanbul), *Planlama* (Special issue on Istanbul), vol. 10, nos. 1-4, pp. 46-51.

Cohen, M. (1995), 'Finding the frontier: Posing the unanswered questions', in I. Serageldin, M. Cohen and K. C. Sivaramakrishnan (eds), 'The Human Face of the Environment', *Environmentally Sustainable Development Proceedings Series no.6*, The World Bank, Washington DC, pp. 276-280.

Davey, K. (1993), 'Elements of urban management', UNDP-UNCHS-World Bank Urban Management Programme, *Urban Management and Municipal Finance Series*, no. 11, Washington DC.

Eroglu, V., Sarikaya, H. S. Sevimli, M. F. (1996), 'Istanbul'daki Sanayi Atiksularinin Denetimi: Degerlendirmeler ve Gelismeler', in I. Talinli, S. Sozen, E. Gorgun and M. Gure (eds), *ITU Insaat Fakultesi, Cevre Muhendisligi Bolumu, 25-27 Eylul 1996* (Proceedings of the ITU 5. Industrial Pollution Control Symposium 1996), Istanbul, pp. 1-28.

Gonullu, R. (1996), 'Draft report on legal and administrative framework for wetlands of Turkey', prepared for Yubarta Inc. for Med-Wet Management Sub-Project, *Legal and Administrative Framework for Mediterranean Wetlands*, edited and published by Ministerio de Medio Ambiente, Dirección General de Conservación de la Naturaleza, Madrid.

Heper, M. (ed) (1985), 'Dilemmas of Decentralization', Research Institute of the Friedrich Ebert Stiftung, *Analysen aus der Abteilung Entwicklungslander-forSchung Series*, no.123-124, Bonn.

Istanbul Metropolitan Municipality (1995), *Annual Report 1995*.

Istanbul Metropolitan Municipality, Department of Planning and Reconstruction, City Planning Administration, (1995), *Istanbul Metropolitan Area Sub-Region Master Plan Report* (1/50,000 scale), Istanbul.

ISO Cevre Subesi (1995), *ISO Cevre Gorusu ve Cevre Calismalari* (Istanbul Chamber of Industry's perspective on the environment and related activities), ISO Publication, Istanbul.

ISO Cevre Subesi (1995), *Proceedings of the Symposium on Environmental Management at the Firm Level (26/10/1995)*, ISO Publications, Istanbul.

Koksal, S. (1990), 'Ticarilesen Gecekondu ve Kent Yoneticileri', *Iktisadi ve Ticari Bilimler Fakultesi Dergisi, Prof. Mubeccel Kiray'a Armagan*, vol. 7, pp. 260-275.

Kurum, Z. E. and Kurum, F. (1991), 'Pollution levels in Omerli reservoir', in *Proceedings of the International Symposium on Management Strategies of Surface Water Resources* (4-6 November), ISKI, Istanbul, pp. 380-404.

Patterson, A. and Theobald, K. (1996), 'Local Agenda 21, Compulsory Competitive Tendering and Local Environmental Practices', *Local Environment*, vol. 1, no. 1, pp. 7-20.

Ribeiro, L. C. de Queiroz and Correa do Lago, L. (1995), 'Restructuring in large Brazilian cities: The centre/periphery model', *International Journal of Urban and Regional Research*, vol. 19, pp. 369-382.

Sen, Z. (ed) (1995), *Proceedings of the Symposium on Water Resources in and around Istanbul*, organised by ISKI and ITU, Istanbul (22-25 May), ISKI Publications, Istanbul.

Sen, Z. (ed) (1996) *Proceedings of the Icmesuyu Symposium,* organised by ISKI, ITU and Su Vakfi (7-10 October), ISKI Publications, Istanbul.

Sonmez, M. (1996), 'Istanbul in the 1990s: A statistical survey', *Istanbul* (Special issue for Habitat II), pp. 43-51.

Stephens, C. (1995), 'Health, poverty, and environment: The nexus', in I. Serageldin, M. Cohen and K. C. Sivaramakrishnan (eds), 'The Human Face of the Environment', *Environmentally Sustainable Development Proceedings Series no.6*, The World Bank, Washington DC, pp. 173-8.

Strong, M. (1995), 'The Road from Rio', in I. Serageldin, M. Cohen and K. C. Sivaramakrishnan (eds), 'The Human Face of the Environment', *Environmentally Sustainable Development Proceedings Series no.6*, The World Bank, Washington DC, pp. 11-15.

Timur, O. and Eren, U. (1991), 'Pollution control in Istanbul's water catchment areas', in I. E. Goknel, D. Orhon, O. Buyuktas, N. Sarpinar, S. Tokta (eds), *Proceedings of the International Symposium on Management Strategies of Surface Water Resources* (4-6 November), ISKI, Istanbul, pp. 455-463.

Tuna, O., Orhon, O. and Bederli, A. (eds) (1991), 'Monitoring of Technology for Pre-treatment of Industrial Wastewater', *ISO-Su Kirlenmesi Arastirmalari Milli Komitesi no. 1* (Istanbul Chamber of Industry, National Committee on Water Pollution Research), Istanbul.

Turkiye Ulusal Komitesi Danisma Kurulu (1996), *Turkiye Ulusal Rapor ve Eylem Plani* (National Report and Plan of Action for Turkey), Report for United Nations Human Settlements Conference, Habitat II, Istanbul.

Uysal, Y. (1994), 'Carpik Kentlesmenin Illegal Yuzu' (The illegal aspect of distorted urbanization), *Istanbul*, no. 11, pp. 71-74.

Uysal, Y. (1995), 'Icme suyu havzalarinda plan, politika ve hukuk' (Plan, politics and law in fresh water reservoirs), *Tesisat Muhendisligi*, nos. 7-8, pp. 51-55.

Varley, A. (1989), 'Settlement, illegality and legalisation: The need for reassessment, in P. Ward (ed), *Corruption, Development and Inequality,* Routledge, London, pp. 156-174.

Yonder, A. (1982), 'Gecekondu policies and the informal land market in Istanbul', *Built Environment,* vol. 8, no. 2, pp. 117-124.

Yonder, A. 1998, 'Implications of double standards in housing policy: Development of informal settlements in Istanbul, Turkey', in E. Fernandes and A. Varley (eds), *Illegal Cities: Law and Urban Change in Developing Countries,* Zed Books, London, pp. 55-68.

24 Research-Management as an Approach to Solving Environmental Conflicts in Metropolitan Areas: A Case Study of the Manizales-Villamaría Conurbation, Colombia

LUZ STELLA VELÁSQUEZ AND MARGARITA PACHECO

Introduction

The incidence of urban environmental conflicts has increased in recent years in Colombia. The enforcement of recent environmental laws and the disparity between the administrative boundaries of municipalities and natural environmental regions have helped exacerbate such conflicts. But a number of local initiatives have also appeared which bring together disparate stakeholders in the search for solutions.

This paper examines the case of a series of environmental conflicts which developed between two neighbouring municipalities in the coffee growing region of Viejo Caldas in western Colombia. The two municipalities are Manizales and Villamaría, both part of a conurbation (i.e. an urban area comprising more than one urban centre, in this case several municipalities), though not yet legally constituted as a metropolitan area in the manner contemplated by Colombian Law.

The successful management of environmental conflict between Manizales and Villamaría has been aided to a large extent by the active involvement of a national network of university-based urban environmental research groups presently comprising 236 researchers. These groups are active in research on local environmental problems and also in providing theoretical and methodological guidelines to local governments and other local actors involved in urban environmental management. They have been instrumental in helping to solve other environmental conflicts of the kind described here.

The Context of Urban Environmental Management in Colombia

Colombia's Political Constitution of 1991 enshrined the right of all citizens to enjoy a healthy environment. In 1993, the Ministry of the Environment and the National Environmental System (SINA) were created. But most importantly, 1993 saw the approval in Parliament of Law 99, known as the Law of the environment, which establishes guidelines and a general framework for environmental management of the national territory.

Since the mid-1980s Colombia's provinces but especially its municipalities have been benefiting from a process of decentralisation of power and resources from central government. This has been positive in that it encourages coordination among municipalities to prepare common policies on key economic and political issues. But this autonomy has been less positive in relation to the environment.

There are no environmental regions for planning purposes in Colombia, which poses enormous problems for the adequate management of the country's highly diverse ecosystems. In practice, for the past three decades most environmental management is carried out by watershed management agencies and by provincial environmental authorities. These have only recently begun to create adequate policy instruments to articulate regional environmental issues with the political and administrative jurisdiction of provinces and municipalities. As a result, the enforcement of laws has brought about numerous conflicts among environmental authorities at the national, regional and municipal levels.

The Geography of the Manizales-Villamaría Conurbation

Manizales and Villamaría are located in the centre of the Andean Region in western Colombia. They are part of the so-called 'Coffee Axis Conurbation', which has historically been a demographically and economically dynamic area of great importance. This was particularly true in the first half of the twentieth century when coffee exports represented the bulk of the country's exports earnings and a sizeable proportion of coffee production came from the region around Manizales known as Viejo Caldas (Velásquez, 1998).

With a joint population of some 438,000 inhabitants in the mid-1990s, the two municipalities share an eco-system and a variety of natural resources, urban and rural infrastructure, public services, transport, economic potentialities and environmental problems in a rural-urban territory of 738 km^2.

The area's mountainous system is characterised by its diversity and a wide variety of natural ecological and geographical units in altitudes ranging from 800 metres to 5,400 metres above sea level. The main source of water in the region is found in a national park in the municipality of Villamaría called Parque de los Nevados. The region is also endowed with considerable hydro-electric potential, while some recent research has been assessing the geo-thermal energy potential from the Nevado del Ruiz volcano.[1]

Both municipalities are located in the Central Andes mountain range, an area marked by frequent geological transformations and tectonic movements as well as considerable volcanic activity. However, of a total of 149 geological events recorded between 1976 and 1992, 122 were caused by different forms of erosive processes and by the inadequate management of watershed areas. The natural environment makes its presence felt with unusual intensity in the two cities.

The conurbation experienced fast demographic and physical growth in the period 1975-1990. This placed strains on the area's natural environment, while a boom in speculative building brought about a physical deterioration of much of the historical core of Manizales. Growth in the five years that ensued slowed down in Manizales but continued apace in Villamaría where population doubled between 1992 and 1997. Many of Villamaría's new developments are located in areas of considerable ecological value, in many cases altering their natural environmental characteristics without due consideration to environmental restrictions and the aesthetic fragility of the landscape.

Urban activities in both cities have been traditionally segregated by land-use zoning practices which have resulted in differential access by diverse social groups to public and community services. Official programmes to rehabilitate the dwellings and the environment in several of the so-called 'marginal districts' of the conurbation have been limited in their effect, so the quality of most housing found there is poor.

Problems have become more acute in recent years due to a rapid increase in population in high risk areas such as steep slopes, making it necessary to invest more resources in infrastructure and other soil stabilisation measures. The number of households in such districts rose from 538 in 1987 to 1,378 (representing approximately 7,000 people) in 1996.

While Manizales is actively putting into practice plans to improve urban and environmental quality in the city's sixteen districts (called *comunas*)[2] with programmes such as drainage, rehabilitation of buildings and construction of recreational infrastructure, Villamaría lacks the financial resources to do so, which is gradually leading to a progressive and rapid deterioration of the municipality's built and natural environment.

For example, in the ten years after 1987 municipally-maintained open spaces and recreational areas in Manizales increased from an average of 2.2 m^2 per inhabitant to 6.3 m^2. The municipality's current development plan has as its objective to create additional recreational areas to reach an average of 9.6 m^2 per inhabitant by 2000 (Municipality of Manizales, 1987, 1990 and 1995). Recreation is one of the main priorities of the plan and a focus of the municipality's environmental policy. By contrast, as a result of the rapid growth of built-up areas, Villamaría has lost a large share of its recreational area which by the mid-1990s was a mere 1.6 m^2 of open space per inhabitant.

The centre of Manizales has considerable environmental potential in terms of the availability of public space use for both local inhabitants and visitors. However, the lack of an adequate programme of revitalisation of the historical centre has meant that a growing number of incompatible activities have sprung up in the city centre, with the consequent deterioration of this important asset. The centre was recently classified as a National Historical Heritage area and included in the Bio-Manizales Environmental Conservation and Urban Renewal Project (Municipality of Manizales, 1995; National University, 1997). This has enabled the introduction of tax incentives to assist owners of buildings with some historical value in their upkeep and restoration.

Recent Environmental Conflicts between Villamaría and Manizales

Urban environmental conflicts between Manizales and Villamaría grew in number and frequency after 1995 when Manizales' Planning Department decided to enforce restrictions to new developments and to the expansion of existing ones. The main sources of conflict have been:

- Environmental problems derived from the extraction of minerals and timber used in housing construction and infrastructure. This had regional and local environmental impacts for which neither municipality was prepared to take responsibility.

- Restrictions imposed by Manizales with a view to minimise the production of solid waste and increase the use of cleaner technologies in industrial production and construction. This has encouraged the location of polluting industries in Villamaría where by-laws and controls are more lax, but has also led to an important reduction in the volume of taxes collected by Manizales.

- Due to its proximity to Manizales and lower land costs, Villamaría has gradually become a dormitory town for commuters who work in Manizales. Planning and building regulations in Villamaría have helped preserve the municipality's comparatively lower land costs which has led to the rapid growth there of low-income housing.

- Tighter controls to the sale of alcoholic beverages for the young and opening hour restrictions for bars and discotheques in Manizales have encouraged the relocation of many of these establishments to Villamaría. One of the effects of these measures has been an increase in the number of traffic accidents along the road connecting the two municipalities.

- The proceeds of petrol sales tax in Manizales are earmarked for the construction of infrastructure and the implementation of the city's transport master plan (Municipality of Manizales, 1992 and 1995).[3] But because Villamaría has a lower tax rate (of 9 percent compared with 15 percent in Manizales) this has had an effect on the economy of Manizales and has reduced the volume of tax collected by 35 percent which has also affected the city's ability to implement a transport plan where preservation of the environment plays a central role.

- A large part of the Los Nevados national park is located within Villamaría's jurisdiction. In spite of the importance of this strategic ecosystem which provides water for both municipalities while also having considerable tourism potential for the region, there are no overall policies for managing the park. And national programmes and projects for areas of natural value developed by the Ministry of the Environment have not been implemented by either municipality (Ministry of the Environment, 1996).

- The river banks shared by both municipalities have been put to very different uses. While Manizales launched a programme of construction of water parks for recreational purposes, Villamaría allowed the location of industries, repair shops and petrol stations there. Joint solutions for solving this particular land-use conflict have been restricted to imposing a few economic restrictions on some of the activities located on the river banks.

As in many other Colombian municipalities, communities in Manizales and Villamaría exercise their right to sue the local authority for failing to protect their environment. Of all the actions filed between 1995 and 1997, 23 percent referred to areas shared between both municipalities. Similarly, 33 percent of citizens' complaints which appeared in two sections of the local newspaper in the same period related to the lack of environmental protection of the conurbated area of Manizales and Villamaría.[4]

An Approach to Solving Inter-municipal Environmental Conflicts

Since 1992 the National Network of Urban Environmental Studies has been actively promoting a process of reflection about urban environmental problems. The network consists of university-based research groups known by their Spanish acronym of GEA-UR. Manizales' GEA-UR became active in 1994 when it co-ordinated the Manizales Urban Environmental Profile with the support of Colciencias, the National Fund for Scientific Research, as a national pilot case-study.

Since 1996, the GEA-URs of Manizales and Villamaría have been studying the urban environmental reality of both municipalities and their conurbation with a view to helping identify solutions for the environmental conflicts described earlier. This work led to the design of a short-term environmental management plan, the Bioplan 1997-2000, aimed at promoting collaboration between the two municipalities. The plan proposes a series of policies and strategies for the development of common programmes and projects aimed at raising the quality of the urban environment and protecting the regional environment across political and administrative divisions.

The drafting of the Bioplan has been seen as a first crucial step towards the solution of inter-municipal environmental conflicts. The process involved a series of phases summarised below.

1. Profiling of styles of environmental management in both municipalities (September 1996)

To assess the viability of any future coordination efforts it was necessary first to understand the management style used by each municipality. This involved a short investigation using secondary information supported by focus group discussions with citizens, NGO staff and municipal officials. The results of the survey were discussed at a first meeting involving the municipal mayors and the regional and provincial environmental authorities. Participants endorse the aim of widening existing knowledge of environmental problems and identifying the potential of environmental and cultural resources of the region with due attention to the social and economics needs of both municipalities. A Technical Inter-municipality Committee was appointed with the task of defining priorities and agreeing on common management styles.

2. Preparation of an Inter-municipal Environmental Profile (November 1996-January 1997)

Following the recommendations of the Technical Committee, and building on the methodological findings of the 1994 environmental profile, the GEA-URs launched a "research-management" process with a view to identifying the main environmental problems and potentialities of the conurbation. This inter-municipal environmental profile was prepared jointly with officials from both municipalities, and researchers of the regional environmental authority and the local branch of the National University. The main aim of this phase was to agree on some form of joint institutional management.

Dissemination of the results motivated other local research groups and individual citizens to participate in the process of environmental management. Regular discussion meetings and citizen forums were set up in both municipalities while a first agenda of inter-municipal priorities was agreed.

3. Preparation of the First Agenda for Joint Action (February 1997)

The list of joint municipal priorities resulting from the environmental profile in phase 2 was complemented with an Agenda of Joint Municipal Action. This set the foundations of a short-term plan for environmental action aimed at mitigating the impact of recent environmental conflicts and establishing mechanisms for developing programmes and projects of common interest as identified in the development plans of both municipalities.

The proposed agenda was discussed with the mayors, the environmental authorities and the Secretaries of Planning of both municipalities. The GEA-URs were given charge of preparing an inter-municipal environmental action plan as well as disseminating information about the process among local and regional stakeholder groups.

4. Creation of an Inter-municipal Environmental Committee (February 1997)

An Inter-municipal Environmental Committee was created with a view to implement the inter-municipal environmental action plan. The Committee's main functions are to oversee the process and to raise resources for the implementation of the short-term plan. It is comprised by the municipal mayors or their delegates, the director of the Corporation for Regional Development (the main environmental authority in the region) or their delegate, the Secretary of Agriculture of the province, the coordinators of the GEA-URs of Manizales and Villamaría, and the representatives of the environmental sector in the Territorial Planning Councils.

5. Coordination with National Environmental Policies (March 1997)

The Ministry of the Environment outlined an approach to urban environmental management based on the principles of participation and the search for urban sustainability. This was the starting point for the integration of the conurbation's inter-municipal action plan to regional plans as well as to policy initiatives of the Ministry. The inter-municipal action plan also became one of a handful of pilot projects in a World Bank-funded programme for the strengthening of urban environmental management in small and medium-sized cities.

6. The Inter-institutional Collaboration Agreement (April 1997)

The process of coordination across institutions at the national, regional, provincial and local levels has been an important contribution of the experience of the Manizales-Villamaría case to the formulation and management of local environmental action plans in Colombia. The search for a joint basis of collaboration was enshrined in an agreement signed by the different environmental authorities. This led to the earmarking of resources to support a joint environmental action plan. Information about the experience was disseminated by the local GEA-URs through the network of environmental researchers in the hope of widening the process to other areas in Colombia.

7. Joint-interest Projects in the Environmental Action Plan

Projects of common interest to both municipalities aimed at seeking swift and effective solutions to inter-municipal environmental conflicts feature in the proposals of the Manizales-Villamaría Bioplan 1997-2000. This short term plan incorporates the aims of the environmental plans of both municipalities into joint programmes and projects to integrate different municipal realities while seeking to dilute the effect of political and administrative divisions. In the plan the environment shared by both municipalities is no longer perceived as two separate realities but rather as a common heritage to be protected jointly. It is recognised that the starting point of both environmental plans is a search for an improvement in the quality of life of their inhabitants.

8. Strategic Projects of the Inter-municipal Bioplan 1997-2000

The Bioplan comprises a set of strategic goals to be achieved jointly in the search for a sustainable development of both municipalities and their conurbation. These have been translated into a number of programmes, including:

- Bio-transport: this proposes amongst other projects a cableway and urban lifts as energy-efficient modes of mass transportation for passengers, serving the more highly populated sections of the conurbation.

- Bio-tourism: the aim is to develop a regional tourism policy which integrates the regional environment into an overall tourism management policy. The plan will give priority to the construction of a "snow route" to be managed jointly by the two municipalities.

- Recreational environmental education: this seeks to incorporate Manizales' water parks and eco-parks into current projects and programmes of recreation and environmental education. This also seeks to give shape to the University of the Environment at the Alcázares-Arenillo Eco-park, located in the municipal boundary between Manizales and Villamaría.

- Joint management of solid waste: this seeks to coordinate the action of existing projects for the recuperation and protection of sources of water in the administrative boundaries between the two municipalities; to minimise the environmental impacts of manufacturing production through joint management and with the participation of the plants located on the banks of the Chinchiná River which marks the boundaries of the two municipalities; and to reaffirm a joint commitment to consolidate and develop a recycling plant currently in operation in Manizales as a pilot project jointly funded by the municipality and a charitable foundation.

Lessons of Experience

In the past few years a series of central government initiatives have sought to create a framework for territorial planning and management. This gives priority to political-administrative divisions over geographical realities while ecological and environmental criteria are increasingly sidelined.

The jurisdiction of Regional Environmental Corporations, the main regional environmental authorities have been increasingly reorganised under political criteria and entrenched practices, with no concern for principles of inter-municipal or inter-provincial environmental management. Adequate watershed management, for example, which is one of the main responsibilities of such Corporations, is made difficult when rivers are used as administrative boundaries between provinces or between municipalities.

These trends must be reversed if we are to seek a greater harmony between the natural and man-made environments. This requires better interaction and coordination among politicians, professionals, and civil society organisations with a view to laying down their interests in favour of a new environmental rationality. Without this, it will be very difficult to give shape to a territorial organisation which promotes sustainable development.

Land-use environmental planning requires adequate knowledge of the existing land uses and the potential uses to which land can be put. Such knowledge must provide the basis for regional planning while also becoming a reference point for local planning. This necessitates an approach to data collection and planning involving wide participation of regional and municipal authorities.

The selection of urban and regional development goals for the region must be agreed among the various social actors, including the state. This involves searching for ways of overcoming conflicts between present and potential land uses in the context of different scenarios, and with well-defined responsibilities and timetables to overcome and prevent such conflicts. This is a major challenge that metropolitan areas and conurbations in Colombia must face in their search for sustainable development.

Notes

1 Research on the area's potential to generate geothermal energy started in 1986 under a joint project between the National University and CHEC, the region's power generation company. A joint venture company, Geonergía Andina SA, was formed in 1992 involving private investors alongside partners like CHEC and a state-owned financial corporation to explore the commercial viability of this initiative. As an enterprise seeking to promote the use of alternative sources of energy Geoenergía Andina benefits from tax incentives and has recently been attempting to attract international investors.

2 A *comuna* refers to a geographical subdivision of the city with a population ranging from 30,000 to 50,000 inhabitants.

3 Since 1986 Colombian municipalities have been entitled to fund the construction and improvement of mass-transit systems from the proceeds of a tax levied on local petrol sales. They are free to set the rate of tax within limits specified by the national government.

4 Information from the 'Denuncie' and 'En línea con el Director' sections of *La Patria* daily newspaper.

References

Corporación Reverdece, 1997, *Informe de Gestión*, Manizales.

Ministry of the Environment, 1996, 'Plan nacional director del Sistema de Parques Nacionales y otras áreas protegidas: Parque de los Nevados', discussion paper (mimeo).

Municipality of Manizales, 1987, *Plan de Desarrollo de Manizales 1987*, Secretaría de Planeación Municipal, Parques y Zonas Verdes, Manizales.

Municipality of Manizales, 1990, *Plan de Desarrollo de Manizales 1991. Manizales Calidad Siglo XXI*, Secretaría de Planeación Municipal, Manizales.

Municipality of Manizales, 1992, *Plan de Transporte Masivo de Manizales*, Secretaría de Planeación, Manizales.

Municipality of Manizales, 1995, *Plan de Desarrollo de Manizales 1995-1997. Manizales Calidad Siglo XXI*, Secretaría de Planeación Municipal, Manizales.

National University, 1997, 'Propuesta para la reglamentación del Centro Histórico de Manizales', Department of Architecture (mimeo).

Velásquez, Luz Stella, 1998, 'Agenda 21, a form of joint environmental management in Manizales, Colombia', *Environment and Urbanization*, vol. 10, no. 2, pp. 9-36.

25 Community-Based Environmental Management in Urban Tanzania

ALPHONCE G. KYESSI

Introduction

In many developing countries, but especially in Sub-Saharan Africa, the growth of population in the urban areas is outstripping the capacity of governments to provide for basic needs such as shelter, water supply and sanitation, drainage, access and other environmental services. In most cases the urban population growth rates have been higher than the corresponding economic growth rates (Mbogua, 1994). Due to the inability of the public sector and the formal private sector to provide affordable housing in suitable locations, large portions of the urban population in developing countries have been forced to build, buy or rent accommodation in unplanned (known also as informal or squatter) settlements (UNCHS, 1994).

Self-reliance and local governance by the poor in their own neighbourhood associations has emerged as a notable phenomenon in many cities of developing countries; left to their own means, the poor have organised to fill in gaps in services left by central and local governments. Among other things, community groups mobilise and organise fund-raising or mutual self-help to provide security, drainage and solid waste management within their immediate area.

Interested parties, including local authorities, should support the efforts of the community organisations in upgrading their neighbourhoods. This is in line with the spirit and concept of the Habitat Agenda, a result of the Second United Nations Conference on Human Settlements held in Istanbul, Turkey in June, 1996. This concept is already being practised in Tanzania and this paper is concerned to document one such case.

Urbanization and Unplanned Settlements in Tanzania

The Urbanization Process

Tanzania, among the poorest Third World countries, is urbanising fast. Between 1948 and 1988 the country had an average urban population growth of 8-10 per cent per annum. The urban population has been doubling every ten years. By the year 2000, it is estimated that, about 50 per cent of the national population will be living in urban locations (URT, 1996). The urban growth gained momentum especially since post independence with the removal of restrictions on local movement within national boundaries. Apart from the rapid urban growth since independence in 1961, urban policies in Tanzania remained fragmented. Urban sprawl and unplanned growth of settlements were the result. As such, provision of basic urban services to settlements such as water supply, power supply, housing, transport and communication, sanitation and drainage has not been able to keep up with the pace of urbanization (Kyessi, 1994; Kironde et al, 1995; Kombe, 1995).

Growth of Unplanned Settlements

Unlike in the case of many other developing countries, the term 'squatter settlements' in the Tanzanian context mainly refers to unplanned residential agglomerations where those who occupy land generally do have quasi legal possession (Kyessi, 1990; Lupala, 1996). The majority of unplanned settlements in the country are a result of either extension of urban boundaries, engulfing former peri-urban village settlements, or development of housing without legal permits from the urban authorities within the urban boundaries. Land is usually subdivided and let out informally in the existing unplanned settlements (Kombe, 1995). Unlike other countries where de-facto occupation and land invasion by squatters is the starting point for squatting, in Tanzania this proceeds by informal purchase of land followed by development without building permits.

Dar es Salaam, the primary and only large city in Tanzania, covers an area of about 1,393 square kilometres. It is estimated that about 70 per cent of the population of the city lives in unplanned housing areas (URT, 1996). There are about 53 such settlements in the city occupying an area of about 50 square kilometres. Failure of the formal planning and management system to control further growth of unplanned settlements has resulted in development of unplanned housing in the so called 'hazardous land'. Unplanned housing has developed within low lying areas usually prone to flooding especially during the rainy season. Examples of such settlements include Hanna Nassif, Msimbazi and Keko. However, even in these areas, with application of appropriate engineering, parts of the settlement could be upgraded to create acceptable living conditions.

Earlier Interventions in Unplanned Settlements Improvement

Unplanned settlement improvement strategies in Tanzania have gone through three distinct stages in the last three and half decades. Between 1961 and 1969 government intervention in housing was chiefly through 'slum clearance programmes'. These were terminated in 1969 due to the fact that this was reducing the volume of the housing stock. Between 1972 and 1980 the focus of government housing policy shifted from slum clearance to a more humane and economical approaches of 'squatter upgrading' and 'sites and services' programmes. With financial assistance from the World Bank substantial physical improvement was achieved in the first and second phases of this programme. The programme being modest in approach and softer in operation than the earlier programme benefited about 600,000 low income inhabitants throughout the country (Kyessi, 1994).

Nevertheless, following the second phase of the programme,evaluations carried out in the early 1980s revealed four major weaknesses:

- there was lack of adequate maintenance programmes following the provision of basic services;
- planning and capital works were carried out by outside contractors without involving the local residents from the neighbourhoods. This created a sense of lack of ownership as the projects were seen as not belonging to them;
- soft loans issued by the (now defunct) Tanzania Housing Bank could not reach the targeted group because of the prohibitive conditions; and
- poor performance of cost recovery destroyed the possibility of replicability of the programme.

A desperate effort by the government to continue this programme into a third phase achieved negligible results. Unplanned settlements listed in the second and third phases, including Hanna Nassif, were never upgraded.

The Emergence of Community Based Organisations in Unplanned Settlements

In the years of the 1990s, we observed the emergence of Community Based Organisations (CBOs) trying to address the increasing housing problems within their community areas which also drew the attention of some multi-national, non-profit organisations towards trying to support these grassroots efforts. The aim of these organisations, as voluntary grassroots groups, has been to operate on the criteria of associations, sharing costs and benefits within a self-defined social or collective interest group (Lupala, 1996).

Unlike the good experiences of such well-known examples as the Kampung Improvement Schemes in Jakarta, Indonesia, the Orangi Pilot Project in Karachi Pakistan and the Community Contracting Method in Sri Lanka, it is important to note that the concept of 'community based upgrading' in providing basic services is rather new in the Tanzanian urban context. As the urban poor are observing that there is very little by way of financial resources available from the government they are therefore moving towards an approach which involves attempting to mobilize whatever local resources that may be available and to use these for the benefit of the community as a whole.

The Rationale of Community Based Organisations for Improving Unplanned Settlements in Tanzania

The period between 1980 and 1990 witnessed consolidation and expansion of unplanned settlements throughout the country with minimum or absolutely no provision of basic infrastructure services. Problems such as seasonal flooding due to lack of storm water drainage, poor accessibility, inadequate social and community facilities, threat to eviction and demolitions have been in most cases the driving force for the formation of CBOs.

In recognition of the possibilities for community based upgrading of unplanned settlements, using CBOs as the basis for self-organisation, a pilot project was established in the unplanned settlement of Hanna Nassif and the remainder of this paper describes and analyses this project.

Community Based Upgrading in Hanna Nassif, Dar es Salaam - A Success Story

Location, Evolution and Size of the Settlement

Hanna Nassif settlement is located in Kinondoni District about 4 kilometres north of the centre of Dar es Salaam. The settlement covers an area of about 50 hectares of land. Its proximity and ease access to the city centre and to higher order community facilities, such as the Muhimbili Referral Hospital, Kariakoo Market and other functions located in the Central Business District (CBD), resulted in higher land values compared to other unplanned settlements within the city.

Before the development of Hanna Nassif as a settlement it was a farm belonging to a person of Greek origin known as Hanna. Upon his death the land was claimed by another settler by the name of Nassif. The area, Hanna Nassif, thus derived its name from these two settlers. When freehold titles were abolished in 1963, the people who were working in the farm found themselves to be classed as squatters overnight. From then on subdivision of individual parcels of unsurveyed land has been going on in the area leading to rapid development of informal housing. In 1975 thgere were less than 1,000 houses with a density of 20 per hectare; by 1994 there were over 2,000 houses and densification continues (Lupala, 1996).

Population, Housing Characteristics and Employment Patterns

The population of Hanna Nassif in the mid 1990s was about 20,000 people with density of approximately 400 persons per hectare, making the settlement one of the most densely populated neighbourhoods in Dar es Salaam. This has increased from some 11,685 in 1975 and 14,567 in 1982 and around 19,000 in 1990. About 60 per cent of the people living in the area are landlords and the rest are tenants.

Hanna Nassif is characterised with Swahili type of housing where rooms have their doors opening to a central corridor. Toilets are outside the house. Current studies in the area reveal that about 73 per cent of the houses are built with permanent materials, 18 per cent semi-permanent and about 9 per cent with temporary materials of mud and pole and roofed with thatch. Commercial activities are also carried out in some of the houses.

Only 27 per cent of heads of households in Hanna Nassif are in full time employment with a further 11 per cent employed part time. Fourteen per cent are self-employed or unemployed. About 7 per cent of heads of households are retired people. Of those employed 50 per cent work in the private sector and 59 per cent of the non-formally employed are active in petty trading (Lupala, 1996).

Community Facilities and Physical Infrastructure

In the early 1990s the state of infrastructure provision in the settlement was rather poor with insufficient community facilities as follows: inadequate drainage meant frequent flooding resulting in insanitary conditions; access was totally inadequate; only 18% of households were connectred to water supply, the rest buying at inflated prices; sanitary arrangements were very poor; many hosueholds lacked electricity connections; solid waste was mainly informally dumpred, buries or burned; and there were no public open spaces.

Searching for Solutions

For many years, the residents of Hanna Nassif were hoping that the government would do more to help them solve their environmental problems; however, this faith proved to be ill-founded. It was not until 1991 that the residents came into collaboration with the city authorities, the International Labour Organisation (ILO) and United Nations Development Programme (UNDP) to come up with an upgrading programme involving the residents from the beginning of plan preparation, through implementation to operation and maintenance. The residents were willing to contribute in cash and in kind to solve their problems, starting with storm water drainage system as priority number one.

Description of the Project

The project aimed at improving living conditions and expanding employment opportunities in Hanna Nassif. In order to generate employment opportunities in the area, the residents started implementing an open storm water drainage system using a community contracting procurement method. The project learned from similar experiences carried out in earlier years in Sri Lanka (UNCHS, 1994). The approach was based on residents being trained to become contractors within their project. This method guaranteed that most of the investment fund would remain in the area thus improving the incomes of the people who are mostly poor. Thus, the principle goal was to build skills of the community, with an essential condition of bringing sustainable development by upgrading local infrastructure, through the generation of employment for the purpose of poverty alleviation among the residents. The project is being carried out in phases.

Organisation

To implement the programme a Community Development Committee (CDC) was formed. The Hanna Nassif CDC is a central body responsible for overall day-to-day operations of community activities in the area. The residents divided their settlement into six zones each electing one member of the CDC. In electing representatives, it was agreed to include women in the CDC and this resulted in 11 members out of 19 being women.

The major role of the CDC is to act as a bridge between the residents of Hanna Nassif and the Dar es Salaam City Council (DCC) and the donor community. In addition, the CDC has responsibility for mobilizing people for project financing, their involvement in infrastructure construction and the future monitoring, operation and maintenance of infrastructure provided by the project.

The CDC has played a vital role of creating community awareness and understanding of their problems through animation and also creating a sense of project ownership and the need for future operation, maintenance and cost recovery. The CDC itself is made up of two subcommittees, one concerned with economic and finance matters and the other of construction.

The CDC is supported by a Technical Support Team (TST) comprising of a programme coordinator, a civil engineer funded by the United Nations Volunteer Service (UNV), a local civil engineer, senior community animator and other technical staff seconded from the DCC to the project, including a town planner, a surveyor, an accountant, an administrator and a store keeper. Other technical support has been made available by the ILO. The overall coordination was provided by UNDP through the Sustainable Dar es Salaam Project. ILO was the executing agency while UNCHS and UNV were associated agencies.

Funds for the Pilot Project

Although the project was spearheaded by the ILO, with the specific focus on employment generation as part of the Urban Infrastructure Works Programme, the on-going city-wide Sustainable Dar Es salaam Project (SDP) necessitated the integration of Hanna Nassif Pilot Project in the city level strategy of the city infrastructure improvement programme. Thus further organisations came to support the Hanna Nassif initiative apart from the support of the ILO. Total funding for the pilot project was eventually over US$ 600,000 with further mechanisms in place to raise local finance and resources in kind.

Lessons of Experience from Hanna Nassif

In order to generate employment and reduce poverty in Hanna Nassif the project adopted a labour intensive approach. This was the first recorded experience in the use of the method in unplanned settlements in urban Tanzania.

Another feature of the Hanna Nassif project is the use of community construction contracts. From the beginning of the project all construction works on roads and drainage were carried out by the residents themselves under the guidance of the Technical Support Team. This approach not only ensured that all the investment funds remained in Hanna Nassif but also introduced the residents to skills needed later for maintenance of the infrastructure.

It is interesting to note that landlords/landladies participated alongside tenants in implementation of the project; this was certainly a breakthrough for urban projects in Tanzania.

Main Tangible Outputs

The following paragraphs list the achievements of the project:

- As of August, 1996 the following works had been completed 600 meters of main drains, 1.5 kilometers of side drains 1.0 kilometres of murram roads. Part of the settlement is being drained during the rainy period via the channels built by the project and this has positively encouraged the residents to become more involved in the second phase.

- About 25,000 worker-days of employment were generated, developing construction skills and demonstrating equality between men and women in the work.

- Road widths were reduced from the official 12.0 to 5.4 metres so that it was not necessary to demolish any houses.

- The CDC now owns a spacious site office which serves as community meeting place and information centre.

- The construction of the roads and drainage has made the residents more cohesive and willing to cooperate than before.

- The Hanna Nassif upgrading model is now being replicated by other CBOs both in planned and unplanned settlements of Dar es Salaam.

- The project has managed to attract more funding from sources other that the original donors i.e funds from the National Income Generating Programme (NIGP) and Ford Foundation which are funding phase two to the tune of about US$ 700,000.

Observed Problems

Four major problems were experienced during implementation of the project as follows:

- Organising community mobilization and participation is a long process which was inadequately anticipated. Information flow from CDC members to the rest of the residents was weak. This affected the community contributions and regular maintenance of the built infrastructure.

- Donor funds were not available at the right time resulting in delays in implementing parts of the project. It was difficult to collect contributions from all the residents of Hanna Nassif. It seems the residents are still in a state of dependence on the DCC as representing the 'great provider of services'.

- There was no clear understanding of the roles and responsibilities of the Local government leaders (Ward Officer, Community

Development Officers, Ward Health Officers etc.) and that of the CDC.

- Lack of experiences with formation, running and leadership of CBOs also contributed to delays in implementing the project.

Conclusions

The culture of involving CBOs in upgrading projects in urban areas is a recent phenomenon in Tanzania. For the past three and half decades of independence, CBOs have not been involved in the upgrading of informal housing settlements mainly because of the Centralized Planned Economy and the fact that the government was the provider. It is no longer so today.

As noted in the foregoing discussion most CBOs lack an adequate financial basis to take off and also they are deficient in managerial and technical capabilities and therefore confronted with some organisational problems in project planning, implementation and maintenance. In view of these weaknesses the need for capacity building of these CBOs to enable them to undertake small scale upgrading works in their localities and how to establish income generating activities is of paramount importance. Capacity building of CBOs should focus on the following aspects:

- *financial mobilisation:*
  both central and local governments should support CBO activities financially especially for projects which aim at improving living environments and generating employment to the local residents.Financial support can be disbursed as annual allocations through local government Budgets and/or from voluntary and other donor agencies within and outside the country. Privately provided services can be used as assets for income generating activities. A credit union could also be one of the initiatives;

- *organisational structure and networking:*
  while there are varying needs in each settlement and community, the organisational structure for CBOs operating in each area should correspond to the needs pursued by each CBO. The local authority should still be seen as the responsible coordinating body of all CBOs in an urban centre. The coordination of CBO activities will ensure

proper planning of technical and financial support by the Local Authorities and also facilitate learning among communities for similar or different project which are being undertaken by other CBOs. This calls for networking of CBOs.

* *technical support:*
  on site training in relevant skills is important in implementing community projects. A Technical Support Team is always necessary. Other skills can also be acquired through short courses organised by formal training institutions. At the same time manuals should be developed based on the experiences from Hanna Nassif.

References

Kironde, J.M.L. and Rugaiganisa, D.A. (1995), 'Urban Land Management, Regularization and Local Development Policies in Tanzania with Special Emphasis on Dar es Salaam', paper presented in a seminar in Abidjan, 21-24 March.

Kombe, W.J. (1995), 'Formal and Informal Land Management in Tanzania, the Case of Dar es Salaam City', PhD Thesis, *Spring Research Series no. 13*, Dortmund University, Germany.

Kyessi, A.G. (1990), 'Urbanization of Fringe Villages and Growth of Squatters, the Case of Dar es Salaam, Tanzania', unpublished MSc thesis, ITC, The Netherlands.

Kyessi, A.G. (1994), 'Squatter Settlements Development and Upgrading in Dar es Salaam, Tanzania', *Journal of Building and Land Development*, vol.3, no.2, Ardhi Institute, Dar es Salaam, pp. 23-30.

Lupala, J.M. (1996), 'Potentials of Community Based Organisations for Sustainable Informal Housing Upgrading in Tanzania. The Case Study of Hanna Nassif Pilot Project, Dar es Salaam City', unpublished MSc thesis, Leuven, Belgium.

Mbogua, J.P. (1994), 'Integrating Environment and Physical Planning', Report submitted to the International Conference on Re-Appraising the City Planning Process as an Instrument for Sustainable Urban Development and Management, Nairobi, Kenya, 3-7 October.

Sheng, Yap Kioe (1983), 'Community participation in the execution of low-income housing projects', in UNCHS (ed), *Reader Composed of Papers Presented at the Workshop on Community Participation*, 17th September to 1st October 1982, UNCHS, Nairobi, Kenya.

Sheuya, S.A. (1997), 'Lessons of Experience from The Sustainable Dar es Salaam Project: Settlements Upgrading', paper presented in the National Consultation

Workshop on the Replication of Sustainable Cities Programme to Include Municipalities other than Dar es Salaam, 12-13 February.

United Republic of Tanzania (URT, 1996), *National Report on Human Settlements Development in Tanzania*, report prepared for Habitat II, Dar es Salaam.

UNCHS (1994), *The Community Construction Contract System in Sri Lanka*, United Nations Centre for Human Settlements, Nairobi.

UNCHS (1996), *Habitat Agenda*, document from the Second United Nations Conference on Human Settlements held in Istanbul, Turkey, 3-14 June.